TELECOMMUNICATIONS
DIGEST

Telecommunications Digest

Herbert A. Pairitz

McGraw-Hill Book Company

New York   St. Louis   San Francisco   Auckland   Bogotá
Guatemala   Hamburg   Johannesburg   Lisbon   London
Madrid   Mexico   Montreal   New Delhi   Panama   Paris
San Juan   São Paulo   Singapore   Sydney   Tokyo   Toronto

A BYTE Book.

1 2 3 4 5 6 7 8 9 10   DOC   DOC   8 9 3 2 1 0 9 8 7 6 5

ISBN 0-07-048097-4{PBK}
ISBN 0-07-048102-4{HC}

LIBRARY OF CONGRESS CATALOGING IN PUBLICATION DATA

Pairitz, Herbert A. (date)
  Telecommunications digest.

  (A Byte book)
  Includes index.
  1. Telecommunication. I. Title. II. Series: Byte
books.
TK5101.P28   1985   384   84-20160
ISBN 0-07-048097-4 (PBK)
ISBN 0-07-048102-4 (HC)

Sponsoring editor: Jeffrey McCartney
Editing supervisor: Margery Luhrs
Book Design: Julian Hamer

TRADEMARKS

Below is a list of trademarks used in *Telecommunications Digest,* along with their corresponding companies.

Teletype—Teletype Corporation
TELEX—Western Union Corporation
TWX—Western Union Corporation
WESTAR—Western Union Corporation
Info-Com—Western Union Corporation
Data-Com—Western Union Corporation
Metro-1—Western Union Corporation
Motorola MC68020 Microprocessor—Motorola Corporation
Spectrum—International Telephone & Telegraph Co.
SPRINT—GTE Sprint Communications
Datadial—GTE Sprint Communications
3270 Dedicated Access Facility—General Telephone Electronics Corporation
Micro-Com Networking Protocol—General Telephone Electronics Corporation
GTE Telenet Interface Program—General Telephone Electronics Corporation
Telenet—General Telephone Electronics Corporation
Telemail—General Telephone Electronics Corporation
WATS—AT&T Communications
Accunet—AT&T Communications
Accunet Reserved 1.5 Service—AT&T Communications
Enhanced Private Communications Service—AT&T Communications
Dataphone Digital Service—AT&T Communications
Data Phone II—AT&T Information Systems
Circuit Switched Digital Capability (CSDC)—AT&T Communications
Information Systems Network—AT&T Information Systems
Net 1000—AT&T Information Systems
Unix Operating System—AT&T Information Systems
Digital Multiplex Interface—AT&T Information Systems
IBM 3270 Format Translation Program—AT&T Information Systems
Centrex—Bell Operating Companies
Telpak—Bell Operating Companies
T1 Carrier—Bell Operating Companies
Data Access Arrangement—Bell Operating Companies
Intelligent Switching System—Infotron Systems Corporation
Rolm CBX II—Rolm Corporation
Ethernet—Xerox Corporation
Autonet—Automatic Data Processing Corporation
Auto-WATS—Automatic Data Processing Corporation
Uninet—United Telecommunications, Inc.
Cylix Communications Network—Cylix Communications Network
Synchronous Data Link Control (SDLC)—IBM Corporation
2701/2703 Transmission Control Unit—IBM Corporation
3705/3725 Communications Controller—IBM Corporation
3276/3278 CRT Display—IBM Corporation
EBCDIC—IBM Corporation

MSNF—IBM Corporation
PEP—IBM Corporation
JCL—IBM Corporation
Network Control Program (NCP)—IBM Corporation
Systems Network Architecture (SNA)—IBM Corporation
Advanced Program-to-Program Communications (APPC)—IBM Corporation
Advanced Communication Function/System Support Program (ACF/SSP)—IBM
  Corporation
3270 Synchronous Communications—IBM Corporation
2780/3780 RJE Stations—IBM Corporation
4331/3081 Computers—IBM Corporation
3274 Control Unit—IBM Corporation
Virtual Machine (VM)—IBM Corporation
Computer Programs (OS/MVS, OS/VS, BTAM, TCAM, ACF/VTAM, CICS, IMS,
TSO, NCCF, NPDA, SSP, NDLM)—IBM Corporation
Intel 8084—Intel Corporation
Tymnet—Tymnet, Inc.
Tymsats—Tymnet, Inc.
X.PC Protocol—Tymnet, Inc.
Digital Express—COMSTAT World Systems Division
ZILOG Model Z-8—Zilog
LightLink—Datapoint Corporation
LMC-418D—Loral Terra Com Corporation
DECNET—Digital Equipment Corporation
MNP Protocol—MICROCOM, INC.

Contents

Preface

The purpose of this book is to provide a sound, basic understanding of both data and voice communications for people who are just starting in the field or who are required to deal with telecommunications and desire a working knowledge of the subject or those who are actively engaged in telecommunications and need a refresher or reference manual. The primary effort has been to present all the information that is needed to deal with telecommunications on a practical day-to-day basis while avoiding specialized material that would complicate the learning process and tend to diminish the interest of readers seeking a broad understanding of the subject.

A special attempt has been made to select the information that is pertinent and is of practical value to the average reader. This selection is based on 23 years of actual experience in the field of telecommunications. It is hoped that this new approach to the subject will result in the reader feeling comfortable about telecommunications and, perhaps for the first time, really understanding it.

I would like to thank Clarence C. Läster, Jr., Associate Professor in the Electronics Department of San Antonio College, for his assistance in editing the book.

Herbert A. Pairitz

TELECOMMUNICATIONS
DIGEST

1

Introduction to Telecommunications

Data communications, or teleprocessing as it is often called, concerns the transmission of intelligible messages and electronic data processing (EDP) information from one location to another. One very common example would be a cathode ray tube (CRT) display-type terminal in St. Louis obtaining current customer order information from a computer database in Chicago. The term telecommunications covers both voice and data communications, which are very closely related. An example of voice communications with which we are all very familiar is the use of the common telephone. Another form of communications, which is referred to as video or television signals, is sometimes intermixed with voice and data communications on the same communications link. Video transmission will be covered in this book, not in any detail, but only as it becomes associated with voice and data transmission. Incidentally, the terms communications link and communications line are often loosely used interchangeably. To be accurate, however, a *link* is a communications channel that connects two stations to each other, and a *channel* is a single path that could consist of either a physical wire connection or a wireless connection (e.g., microwave radio-frequency transmissions). The term communications line usually refers to a physical wire or cable-type connection only, and the line could be divided into multiple individual communications paths, or channels.

Data communications is based on the transmission of low-voltage pulse-type signals that have discrete meanings, namely, either a 1 (ON) or a 0 (OFF). Since these two digits comprise the entire basic structure of data communications information transfer, it can be seen why the reference is to "digital" communications. At first glance, it would appear impractical to represent all of the common forms of information

1

by using only binary 1 and 0. However, when we look further, we can see that combinations of 1s and 0s in predetermined value positions can define both numbers and letters (e.g., 0011 = 3, and 1100 0001 = A). Also, dots, lines, etc., can by proper positioning on a matrix paper or display screen, be used to draw pictures and diagrams from digital input.

Just as a telephone conversation requires a communications line or path with a telephone instrument at each end, a data communications session requires a communications line with a data terminal (or computer) at each end. Data transmission is further complicated by the fact that it most often uses the same communications line that was originally designed for only voice communications. Other equipment (e.g., modems) is needed to get around the problem. The connection of multiple communications lines to a large computer results in still further complications, which will be discussed in Chapter 6.

Voice and Data Communications Relationship

Our voice telephone system has been around for a long time; it preceded the transmission of computer data-type information by many years. Data transmission was designed to utilize existing voice communications facilities that were made available by the Bell System and by independent telephone companies throughout the United States. Both voice and data communications are concerned with the movement of information from one location to another, but the forms in which the information is entered and delivered are quite different: one form is audible tone conversation, and the other is digital signals that can be interpreted only by a data terminal or computer.

The voice networks constructed by the various telephone companies were originally designed to handle the audio type of signal, referred to as *analog,* the frequency and amplitude of which are varied in direct proportion to the tone and loudness of a person's voice. Computer-type data, on the other hand, requires a communications path that will accommodate a stream of ON and OFF *digital* pulses, called *bits,* that are interpreted by a data terminal or computer and translated into characters of alphanumeric information.

As the need for data (digital) communications increased, the telephone companies devised digital communications lines that were designed to handle the data signals more effectively than analog lines. Though most of the nation's data is still being transmitted by using the older, analog lines, the balance is shifting toward the newer, digital lines. It was discovered that the larger-capacity digital lines (T-1 Carriers) were very effective for providing 24 or more voice channels on a single line. Now the situation has been reversed: the voice signals, in order to use a digital line, must be modified or digitized by a procedure called *pulse code modulation* (PCM).

We can see voice and data communications are closely related, since they can use each other's communications lines. The relationship becomes even closer when we realize that both voice (digitized) and data signals can be transmitted together on the same digital communications channel (T-1 Carrier) and can be accommodated by the same digital telephone switchboard or *private branch exchange* (PBX).

Elements of Data Communications

The basic elements that make up a data communications network are generally referred to as *data terminal equipment* (DTE), *data communications* (or circuit terminating) *equipment* (DCE), and the communications channel or medium. The DTE operates in a digital mode and includes such equipment as CRT terminals, printers, and/or computers. The DCE converts the digital signals to a form, usually analog, acceptable to the communications channel and includes such equipment as modems and multiplexers. The communications channel or medium can be a physical telephone line, microwave channel, or optical path. Figure 1-1 shows a typical data communications network, and a description of each basic element is provided below.

Display Terminal—A means of keying in and viewing data; it also contains a communications control board (see Fig. 1-2).

Modem—A box containing electronic logic boards to act as an interface between the digital (square-wave) terminal and the analog (sine-wave) telephone network (see Fig. 1-3).

Phone—A regular handset for dialing a call and making a connection.

Phone Line—Normal line that could also be used for a voice telephone call.

Phone Company Central Office—Local telephone company switching center closest to the user's location (see Fig. 1-4).

Front End Processor (FEP)—Microprocessor box (computer) that handles the communications lines for the mainframe computer (see Fig. 1-5).

Host Computer—Centralized mainframe computer that (1) contains the business application programs and disk files for the data and (2) controls the communications lines via the front end processor (see Fig. 1-6).

Not all of the elements listed above are contained in every data communications network. For instance, smaller computers often do not require the use of a separate front end processor but do have a

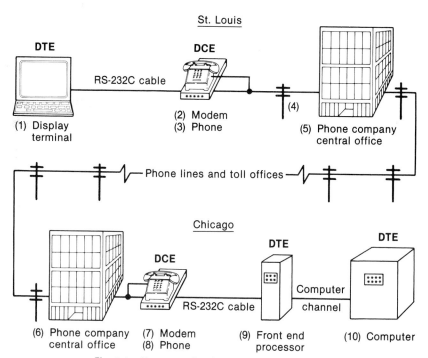

Fig. 1-1 Elements of a data transmission network

Fig. 1-2 CRT display terminal (*Teletype Corp.*)

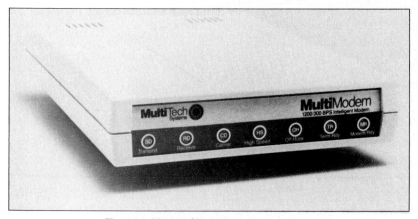

Fig. 1-3 Modem (*Multi-Tech Systems, Inc.*)

Fig. 1-4 Telephone company central office equipment (*Illinois Bell*)

Fig. 1-5 Front end processor (*IBM Corp.*)

Fig. 1-6 Host computer (*IBM Corp.*)

limited capacity for the same basic functions built right into the computer itself.

Sequence of Events in Data Transmission

The events that take place in data transmission are described below; refer to the elements shown in Fig. 1-1.

- When the display terminal operator in St. Louis wants to use the system, he or she brings up the terminal (1) and then dials the telephone number (3) of the computer facility (10) in Chicago. The computer facility is probably set up to answer the call automatically; it notifies the modem (7) in Chicago to send a high-pitched tone to the telephone (3) in St. Louis, where the terminal operator, on hearing the tone, presses a button on the phone to "go to data." This cuts the phone off the communications line and connects the modem (2) at his or her end.

* The display operator in St. Louis now goes to the terminal keyboard (1) and keys in a coded command to request the use of a communications port on the front end processor (9) in Chicago. After the terminal operator's request has been verified for security and accepted by the company facility in Chicago, a communications path from the St. Louis terminal to the Chicago computer is set up and reserved for that particular terminal. (A switched, or dial-up, connection usually requires the transmission of an identification code by the remote terminal.)

* The data being keyed into the terminal (1) at St. Louis is sent on its way, perhaps a character at a time on less sophisticated terminals like teletypes, or accumulated in a buffer for a group transmission on the more sophisticated terminals like most CRT displays.

* In order to transmit data signals back and forth there must be a method of regulating the traffic. The modem must be notified that the terminal operator wants to send data; the computer must be ready to receive the data and not erroneously attempt to send data of its own back to the terminal at the same time; and so forth. This traffic regulation is accomplished primarily by combining two methods. The first depends on a scheme of low-voltage signals that are turned ON and OFF by the terminal and modem to indicate such conditions as DATA TERMINAL READY, and REQUEST TO SEND. These signals are sent on a short cable that connects the terminal to the modem. The RS-232C interface employs a 25-pin arrangement that has been used quite extensively as a standard in this country. The second method of traffic regulation depends on a *communication line protocol* whereby special data characters are transmitted to indicate such necessary functions as BID FOR LINE (ENQ), START OF TEXT (STX), and END OF TRANSMISSION (EOT).

* When the terminal operator (1) completes entering data on the screen, he or she presses the ENTER key to send the data out on the communications line. The communications control portion of the terminal (sometimes this is a separate control unit that

services several displays) adds the required control characters in accordance with its programmed logic circuit board, which contains the line protocol and modem interface voltage requirements.

- The modem (2) passes the control characters and text data (operator's request) onto the communications line. Most telephone lines, particularly the ones you dial up, handle only analog (tone-type) signals, so the main purpose of the modem is to "modulate" a flat carrier signal in accordance with the ON and OFF bit signals supplied by the digital data terminal. This modulation by the modem can be accomplished by varying either the amplitude, frequency, or phase to depict the original digital ON and OFF square-wave signals. The modulation techniques will be covered later in Chapter 4.

- The now-modulated signal goes out onto the telephone network lines (4) to the nearest telephone company central office switching center (5) for routing across country through other toll office switching centers until it arrives at the central office (6) in Chicago closest to the computer facility location and then on into the modem (7) at that location.

- The modem (7) at the Chicago computer facility performs the reverse of what was done by the St. Louis modem. It demodulates the analog signal back into a digital square-wave signal before presenting it to the front end processor (9).

- The front end processor (9) functions as an interface between the host (mainframe) computer (10) and the communications line. It receives the transmission signal bits and assembles them into characters coded as required by the computer (usually Extended Binary-Coded Decimal Interchange Code, or EBCDIC). The front end processor also has many other functions to be discussed in Chapter 6.

- The host computer (10) is, of course, the source of all previously stored information concerning a data system. The information is stored on disk, tape, or even in the computer memory, and it is selected and manipulated by a system application program operating in the computer memory. It would be impractical for each system application program to contain all of the instructions needed for data communications, so a separate common communications monitor program, like IBM's Customer Information Control System (CICS), is employed. It utilizes an access method program, like IBM's Basic Telecommunications Access Method (BTAM), for the detailed work required to exchange data and control characters with the front end processor (9). Thus when the computer receives an interrupt flag signal from the front end processor indicating that data is available, it accesses that data a character or more at a time by using the access method software (e.g., BTAM) under the supervision of the communications monitor software (e.g., CICS). It supplies the data to the application software (e.g., an electronic mail system). The procedure for the computer to send data to the terminal in St. Louis is basically the reverse of what we have just gone through.

Capacity of a Communications Line

The capacity of a communications line to convey information is limited by the line's bandwidth and electrical noise characteristics. Bandwidth, measured in hertz (Hz) or cycles per second (cps), is determined by the range of frequencies the communications line can transmit. In general, greater bandwidth and lower noise characteristics result in a higher communications capacity. For example, short-range telephone lines employing twisted pairs can handle bandwidths of 4000 Hz or higher. On the other hand, standard long-distance telephone channels are limited to a bandwidth of about 3000 Hz (or between 300 and 3400 Hz). Coaxial cable and microwave communications are capable of wide bandwidths of 6 MHz (6 megahertz, or 6,000,000 Hz) and higher.

Electrical noise in communications lines includes internal noise generated by the communications equipment and external noise transmitted into the communications channel. Either type of electrical noise can reduce the line's signal-to-noise ratio and degrade the line's capacity.

The theoretical capacity of a communications line or channel is a function primarily of frequency bandwith (hertz, or cycles per second) and secondarily of signal-to-noise ratio. This is defined by *Shannon's law,* which reads

$$C = W \times \log_2 (1 + S/N)$$

where C = maximum capacity
     W = bandwidth, in hertz
     S = signal power, in watts
     N = noise power, in watts

For a 3000-Hz voice-grade channel the maximum capacity equates to 20 to 30,000 bits per second (bps), depending on the actual signal-to-noise ratio. The theoretical capacity, however, is never achieved because of the excessive cost of near-perfect modems, lines, etc. It can be seen that the exclusive use of a pair of wires rather than a 3000-Hz channel leased from the telephone company can provide a higher bit transmission rate, since a greater bandwidth is available.

Analog communications lines are usually specified in terms of bandwidth, in hertz, or the number of voice channels the line will handle. On the other hand, data communications line capacity is almost always expressed in bits per second, or bps. A standard telephone line used for data communications can handle 9600 bps and up. The T-1 Carrier, a special engineered digital line, can handle up to 1.544 Mbps (million bits per second). Digital voice (i.e., analog voice signals converted to digital signals) is often transmitted via digital communications lines—each voice channel requires about 64,000 bps. Thus a T-1 Carrier can handle up to 24 voice channels simultaneously.

The transmission of data via a communications network is almost always handled on a serial or bit-by-bit basis. Thus eight binary bits, or 1s and 0s, representing a character of information (such as the letter E or the number 9) will be transmitted one bit at a time. This approach is logical, since there is usually only one communications path to the destination. This results in a slower transmission speed than that enjoyed by the various peripherals (printers, disk drives, etc.) of a computer system. In a computer system the data is transmitted in a parallel mode and the lines or paths connecting the peripherals to the computer system require a minimum of eight wires, one for each bit.

2

Communications Media

A variety of methods can be employed to transport data signals from one location to another. Though regular Bell System telephone lines are still the most popular means for transmitting data, some of the alternatives are rapidly catching up. A brief description of each of the media follows.

Dial-up Phone Lines

The dial-up or switched telephone line is the medium with which we are all very familiar, and it was initially used only for voice conversations. You can place a telephone call to anywhere in the United States or the world and talk to a friend or place of business. On the very same call you could send data using a TWX-type teletypewriter with an acoustical coupler. The point being made here is that the same old telephone network can be used for data transmission with little or no modification. The Bell System does require some protection against faulty connections that could interfere with its network. Either a separate Data Access Arrangement box must be installed or the modem or terminal must contain an approved comparable isolation circuit.

ADVANTAGES:
- There is little or no additional cash outlay or installation delay to contend with.
- The costs are completely variable in that you pay for only the time actually used, and thus the costs can be lowest for low-volume usage.
- The Bell System is the most convenient and economical temporary or trial arrangement.

- With portable terminals and acoustical couplers, every telephone in the country, or the world practically, becomes a prospective transmission or receiving location. (Even the touch-tone phone itself can become a data entry terminal.)

DISADVANTAGES:

- As the volume of activity increases to something like an hour or two a day, this alternative starts to become the more expensive approach.
- Unlike leased lines, dial-up lines cannot be conditioned to provide better operation. "You take what you get," and resolving cross-country line problems becomes involved because of the multiple routes. Of course, there is the option of redialing and trying another route if you get a bad connection, but the lines are still inferior to leased lines in many respects.
- Dial-up lines are only two-wire, so four-wire full duplex operation provided by some synchronous protocols is not available. (A limited two-wire full duplex operation that is available is discussed in Chapter 4.) The speed of dial-up lines is normally limited to 4800 bps but may be increased to 9600 bps or even higher by the use of specially constructed modems.
- Dial-up lines are restricted to the point-to-point type of operation and are not capable of a polling or multidrop operation in which all locations must be connected at all times.
- The telephone company has the option of routing the dial-up calls via microwave or satellite line at any time. The latter could present some problems involving propagation delay (a half second round trip) and modem incompatibility (connection failures).
- It takes a half minute or so to make a dial-up connection, which becomes significant when transmissions are multiple and short.

Analog Leased Lines

The general definition of a leased, or private, communications line is a line that is used exclusively by one customer or shared by a group of customers but is not available to the general public on a switching arrangement. The line is usually set up permanently on a full 24 hour per day basis for a fixed monthly rental fee plus a one-time installation charge. This can be arranged for with the Bell System or one of the other common carriers within the United States or around the world. If a line from New York to San Francisco is leased, drops can be added for a nominal charge at, say, Chicago and Denver along the route and, for a more sizable charge at, say, Los Angeles and Memphis, which are not on the direct route. Leased lines become less expensive as the length per using location decreases. The lines go through the phone company's switching centers, but they are not normally affected by the switching mechanisms themselves.

Another type of leased line that is available is the *unloaded short line*. It does not contain coils or repeaters, and so it can be used to transmit digital square-wave signals by means of such devices as the less-expensive limited-distance modems. The Bell System offers unloaded lines only within the same central switching office, which is called 3010 circuits. Of course, if you string your own wire within a building or a complex, you will have an unloaded line.

A new concept of the term leased line is currently evolving. The installation of more advanced network switching equipment by the common carriers has made it feasible to provide a customer with a ''leased line'' that will be available whenever the customer requires it but can, in part or almost entirely, be used by other customers when not needed by the leasing customer. Dialing is not required to access the line, and yet the physical wiring, etc., is not dedicated to a specific customer. This new ''virtual private network'' permits the carrier to utilize its line facilities more fully, and a reduction in costs to the sharing customers results.

ADVANTAGES:

- Leased lines are normally of better grade than dial-up lines. Since the same route is taken each time, problems can be resolved more easily and line conditioning can be added. Several different types of line conditioning (e.g., C1, C2, and D1), available at a nominal charge, are suggested by different modem manufacturers to improve the communications capability of the line.
- Leased lines permit higher transmission rates than the dial-ups. The latter are normally restricted to about 4800 bps and the

former to about 9600 bps, although these speeds can be exceeded through the use of specially constructed high-speed modems.

- In higher-volume and multipoint applications the leased line can be markedly more economical.
- The use of a leased line permits the "polling" of several locations via a single port on the front end processor.
- With a leased line the terminal operator saves the half minute or so that it would take to make a connection with a dial-up line. This becomes almost a requirement if a terminal is to be used for short periods throughout the day.

DISADVANTAGES:
- Leased lines require a longer lead time to install. There is a fixed monthly rental fee and a one-time installation charge.
- It may not be economical for low-volume locations or even for higher-volume locations that are widely separated.

Digital Leased Lines

The digital line is a fairly recent innovation of the Bell System which permits the transmission of square-wave digital signals over long distances. Instead of being mere amplifiers that strengthen the signals at frequent intervals, the digital line repeaters actually recreate the square-wave signals and remove rather than amplify any distortion. As a result, 56,000 bps can readily be accommodated on a four-wire leased phone line, and 9600-, 4800-, and 2400-bps speeds also are available.

A modem is not needed with digital transmission, but a data service unit, which looks like a modem, is needed. It does not modulate the signal, but it does perform such functions as decoding, timing, synchronous sampling, and recognizing and generating control signals. The data service unit can be leased or purchased from a vendor other than the common carrier (Bell System). A channel service unit, which is the more expensive of the two devices required, may have to be leased from the common carrier because vendor offerings of this item are not prevalent. AT&T refers to its digital service as Dataphone Digital Service (DDS).

ADVANTAGES:

- As would be expected, actual experience has indicated that the digital leased lines are more reliable than the older analog-type lines. To make its digital lines even more reliable, AT&T has instituted a plan of alternate routing between phone company switching centers in the event of digital trunk failure. (Of course, the section of line between the data station and the nearest telephone company central office cannot be rerouted, so it remains the weakest point on the entire line.)
- All points on a digital line can be checked out by the AT&T central test location while you are on the telephone.
- A digital line affords a higher speed, say, 56,000 bps versus 9600 bps or so with an analog line.

DISADVANTAGES:

- In most cases a digital leased line would cost more than an analog line, but it may be worth the difference. The digital line appears to be priced so that it costs about the same as the analog line with the two modems. However, there are variations, and quotations should be obtained for both and then compared. The digital line costs less for an A city than for a B city, so a digital line to all A cities could be more competitive.
- Digital lines are not available in all cities. It is possible to add an analog leg to a multidrop digital line, but the cost increases and there could be compatibility problems.
- One of the drawbacks of the initial offering of DDS is that the user cannot run diagnostic tests over the circuit and normally is denied use of the circuit while AT&T performs its own testing to resolve a problem. A new version of the service, called DDS 2, is expected to become available in 1985. It provides a secondary channel that will be useful to the user who wants to transmit diagnostics while the circuit is in operation.

Digital Dial-up Lines

AT&T plans to offer a high-capacity, full-duplex digital dial-up line service by about the end of 1985. Known as circuit switched digital capability (CSDC), this new service will feature a digital transmission rate of 56,000 bps for data, high-quality voice, graphics, or facsimile information. The user data terminal will be compatible with RS-232C and other standard configurations. Virtually error-free operation will permit economical transmissions of high volumes of data or even computer programs in a most timely manner. For example, high-resolution facsimile transmissions will require only about 4 seconds (s) per page via the CSDC line.

The new CSDC system is made possible by previous upgrades of the Bell System, namely, the electronic switching systems (ESS) and the T-Carrier digital lines. A time compression multiplexing scheme is used in CSDC to transmit the 56 kbps (kilobits per second) via the 1.544-Mbps T-Carrier line in short bursts of data. The 1.544-Mbps data bursts are then reconstructed as a continuous 56-kbps bit stream at the destination point. The user CSDC terminal, to be leased from the local telephone company, provides four wires for full duplex data transmissions, two wires for voice communications, and two wires for switching between voice and data modes of operation. This terminal equipment will be compatible with RS-232C and other standard configurations.

Microwave Communications

Microwave radio communications systems employ transmitter and receiver terminals operating at frequencies of about 1 to 30 GHz (1 GHz is 1,000,000,000 Hz, or 1 billion hertz). These very high frequencies, along with highly directive antennas, or "dishes," provide very narrow beams of radio waves for reliable line-of-sight communications. Thus, repeater or relay stations are required at about 40-mile (mi) intervals along the communications link. This method of transmitting voice, data, and video signals is employed by the Bell System and other common carriers for extensions to both dial-up and leased lines, often without the knowledge of the user of the network. Microwave terminals can be economically justified by smaller users when large volumes of communications are required over relatively short distances.

The very high frequencies involved in microwave operation allow for low power levels and high communications bandwidths. Analog microwave radio terminals typically handle up to 1800 voice channels or one television signal on a 6-MHz bandwidth. Microwave digital radio terminals are available for transmission of digital voice, data, or video signals. A typical microwave digital radio operating at 10.7 to 11.7 GHz can transmit up to 1344 digital voice channels at 90 Mbps with a 40-MHz bandwidth. Another means of transmitting radio waves that is much less popular concerns "wave guides," which consist of metal tubes through which the radio waves are transmitted.

ADVANTAGES:

- Microwave communications systems provide for high capacity and highly reliable communications of voice, data, or video information. The entire system can be controlled by the user without relying on the telephone company's local loops.
- Microwave terminals require very small output power levels. Analog microwave radios typically operate at power output levels of about 1 watt (W) or less. Wideband digital microwave radios require output levels of about 5 W.
- Microwave communications can be economically advantageous, especially where high volumes of transmissions, both voice and data, are required for short distances. Longer communications paths, which require construction of relay stations, would have to be economically justifiable to the user.

DISADVANTAGES:

- Microwave radio terminals and relay stations are expensive in terms of both initial procurement and maintenance. Lead times involved in procuring equipment, setting up relay station facilities, and obtaining the Federal Communications Commission (FCC) license for the microwave stations must all be considered before a corporate decision is made to implement a microwave communications system.
- Microwave communications are sometimes degraded by weather conditions, particularly rain, or the intervention of any object in the line-of-sight path of the microwave beam.

Figure 2-1a,b shows a typical microwave transmission antenna dish and its associated transmitter and receiver equipment cabinet. The Model LMC-418D shown is offered by Loral Terra Com Corp., and it is the type of system that could be installed by a company with some

(a) Antenna dish

Fig. 2-1 Microwave system (*Loral Terra Com Corp.*)

fairly high volume communications traffic, perhaps between two business locations 30 mi apart. The digital microwave radio has a 20-MHz radio-frequency (RF) bandwidth and can operate at a speed of up to 26 Mbps. The single transmission stream can be divided into multiple lower-speed communications channels by means of a digital multiplexer. For instance, a 19.062-Mbps stream can be multiplexed into 12 standard DS-1 channels (T-1 equivalent of 1.544 Mbps each).

Optical or Light Communications Systems

One innovative data communications system, similar to microwave systems, utilizes high-capacity light-wave transmissions for short distances. For example, the need to pass high-volume and/or high-speed

(b) Equipment cabinet

Fig. 2-1 (*continued*)

data between buildings of a complex often arises. Microwave terminals could be used for such an application, but the lead time involved in obtaining FCC approval for a microwave license and allocation of non-interfering transmission frequencies may make the light terminal an attractive alternative. No FCC permit is required for the light-wave terminals. The source of light was originally infrared light emitting diodes, but the search for higher speeds and increased optical power has led to the use of solid-state lasers.

LightLink is typical of the light-wave terminals designed for data communications applications. Functional characteristics for this system include full duplex operation at data rates up to 2.5 Mbps with a range of about a mile. The system employs infrared light waves as the transmission medium, and a power output of only 4 milliwatts (mW) is adequate for reliable communications.

ADVANTAGES:
- Light-wave communications offers an economical means of transmitting large volumes of data over short distances. These communications terminals may be installed on roof tops or be adjacent to windows.
- No FCC licensing is involved in the installation and operation of this type of data communications system.

DISADVANTAGES:
- As in microwave communications systems, light-wave communications systems may be impaired by severe weather conditions or the intervention of any object in the line-of-sight path of the light waves.

Satellite Communications

Satellite communications systems employ microwave terminals on satellites and ground or earth stations for highly reliable and high-capacity communications circuits. The communications satellites are positioned in geosynchronous orbits about 22,000 mi above the earth. Thus the rotation of the satellite matches that of the earth, and the satellite appears motionless above earth stations. Three equally spaced satellites are required to cover the entire world (see Fig. 2-2a,b).

The satellite's microwave terminals receive microwave signals

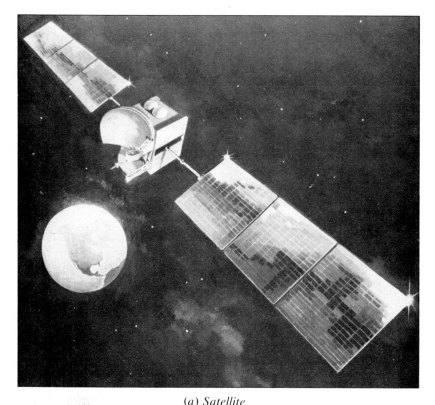

(a) Satellite

Fig. 2-2 Satellite system (a, GTE Spacenet Corp.; b, Harris Corp.)

from an earth station and retransmit those signals on another frequency to another earth station. Because of the long distances involved, the round-trip communications path takes about a half second. This is referred to as the *propagation delay*. The propagation delay on a regular terrestrial phone line is about 1 millisecond (ms) per 100 mi (see Fig. 2-3).

Each microwave terminal on the satellite, designated as a repeater or transponder, includes a receiver for up-link transmissions and a transmitter for down-link transmissions. Separate bands of frequencies for up-link and down-link transmissions are designated in the 1.5–30 GHz frequency range (1.5 GHz is equal to 1,500,000,000 Hz, or 1.5 billion hertz). Typical frequencies for communications satellites are 4–6 GHz for INTELSAT 5 and 12–14 GHz for Anik-B, a Canadian satellite.

(b) Earth station dish
Fig. 2-2 *(continued)*

Each satellite transponder typically has twelve 36-MHz channels which can be used for voice, data, or television signals. Early communications satellites had some 12 to 20 transponders, and the later satellites have up to 27 or more transponders. INTELSAT 5, for example, has a total of 27 transponders. A typical configuration includes 25 transponders providing 24,500 data/voice channels, one transponder providing two 17.5-MHz TV channels, and one SPADE transponder with 800 channels. SPADE (an acronym for *s*ingle carrier per channel, *p*ulse code modulation, multiple *a*ccess, *de*mand assignment) is a digital telephone service which reserves a pool of channels in the satellite

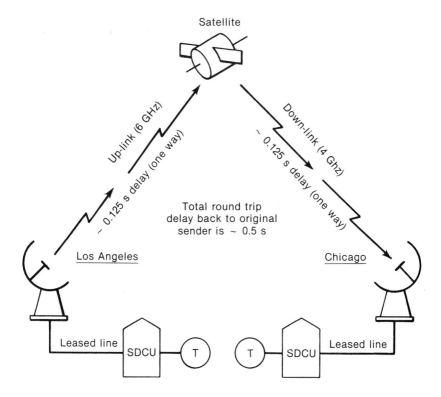

SDCU = Satellite delay compensating unit (where required)
  T = Terminal

Fig. 2-3 Satellite system components

for use on a demand-assignment basis. SPADE circuits can be activated on a demand basis between different countries and used for long
or short periods of time as needed.

PROPAGATION DELAY

The approximate quarter second one-way propagation delay in
satellite communications affects both voice telephone and data communications. Users of voice telephone communications via satellite
links face two objectionable characteristics: delayed speech and return
echoes. Echo suppressors are installed to reduce the return echoes to
an acceptable level. When this problem is cleared up, telephone users
find it relatively easy to ignore the delay characteristics.

Data communications operations face more serious problems caused by propagation delay. Line protocol and error detection/correction schemes are slowed down dramatically by the quarter second of delay. User response time requirements can be difficult to meet because of these cumulative effects.

Satellite delay compensation units are available to ensure a connection and afford better operation for the terrestrial communications terminals that were never designed to deal with the propagation delay of communications satellites. One delay compensation unit is required at each final destination. The units reformat the data into larger *effective transmission blocks* so that retransmission requests are sent back less frequently. This reduces the number of line turnarounds, each of which requires about a quarter second to go from or return to the destination terminal or computer. One error detection and correction method used, called GO-BACK-N, requires that all blocks of data held in the transmitting buffer, back to the one with the error in it, must be retransmitted. A more efficient method is to retransmit only the block of data with the error, but this requires more logic in the equipment at each end.

LINK TO EARTH STATION

Most users cannot afford a satellite earth station, so a land line (leased phone line usually up to 65,000 bps) is needed for a connection to the nearest earth station. Because of the great distance the signal must travel in space, the relatively short distance between the two users on earth becomes insignificant and actually does not affect the operating cost. It is generally not economical to use satellite communications for distances less than about 500 mi, but beyond that distance satellite communications become very economical. This is particularly true of high-capacity or broadband applications. Even though operating costs are insensitive to distance, satellite companies may still charge more for longer distances based on terrestrial line competition.

NONTERRESTRIAL PROBLEMS

The nonterrestrial portion of satellite communications bypasses the problems encountered with broken phone lines, etc., but it has its own unique set of problems. Since satellite communications employ high-frequency microwave radio transmission, careful planning is required to avoid interference between the satellite and other microwave systems. Eclipses of the sun, and even the moon, can cause trouble because they cut off the source of energy for the satellite's solar batteries. Backup batteries are used to resolve most of these difficulties, but the problem that is the most severe is when the sun gets directly behind the satellite and becomes a source of unacceptable noise. This occurs 10 times a year for about 10 min each time. In order to obtain uninterrupted service, an earth station must have a second dish antenna a short distance away or the single dish antenna must have access to another satellite.

ACCESSING THE SATELLITE

There are three methods by which multiple users (earth stations) can access the satellite. The first is frequency-division multiple access (FDMA), whereby the total bandwidth is divided into separate frequency channels assigned to the users. Each user has a channel, which could remain idle if that user had no traffic. Time-division multiple access (TDMA) provides each user with a particular time slot or multiple time slots. Here the channels are shared, but some time slots could be idle if a user has no traffic to offer. With code-division multiple access (CDMA) each user can utilize the full bandwidth at any time by employing a unique code to identify the user's traffic. There are, of course, trade-offs among the three methods; they involve error rate, block size, throughput, interference, and cost.

ERROR CORRECTION

The high propagation delay time of approximately 125 ms one way makes it necessary to be very selective in the specifications for the error correction method. Frequent line turnarounds for acknowledgment of short blocks of data would slow up the total throughput considerably and could render the use of satellite transmission impractical. If stop-and-wait automatic request for repeat (ARQ) is employed, the receiving station, upon detecting an error, must notify the originating station so the erroneous data block or frame can be retransmitted. If the operation is half duplex, this requires a line turnaround with the associated propagation delay. Full duplex operation would be much more efficient in this situation.

Continuous ARQ permits the originating station to transmit a number of frames before the receiving station must send an acknowledgment or error message. At the time of writing this was seven frames for Synchronous Data Link Control (SDLC) protocol and 128 for high-level data link control (HDLC) and some others, but increasing SDLC to 128 was being given serious consideration. Channel throughput is subject to the correct balance of transmission speed, frame size, and number of frames to be transmitted without acknowledgment.

ADVANTAGES:
- Satellite lines are exceptionally well suited for broadband applications such as voice, television, and picture-phone, and the quality of transmission is high.
- Satellite lines are generally less expensive for all voice and data types of transmission, whether it be dial-up or a leased line that is not short. This is particularly true of overseas transmissions, and there is no underwater cable to create maintenance problems.

DISADVANTAGES:
- The propagation delay of about a quarter second one way requires the participants of a voice conversation to slightly delay their responses to make sure no more conversation is still on the way. With a little practice one can get accustomed to the delay, but it does extend the call over a half second each time there is a conversation turnaround. The propagation delay has a more severe effect on the transmission of data, and the effect becomes more pronounced with higher speeds, half duplex operation, smaller blocks of data, and polling. Satellite delay units, front end processors, multiplexers, and other devices have been designed to get around these problems, but there is no solution to the half second lost in total response time for interactive applications. The frequent line turnarounds for error block checking under bisynchronous protocol results in inefficient satellite communications. The newer bit-oriented protocols such as SDLC get around this problem by delaying and reducing check count responses.
- Some of the modems currently in use have not been designed to handle the long delay of the initial connection via satellite, and

the result can be a lost connection. This can be frustrating when the common carrier elects to use satellite lines for regular dial-up calls up to, say, 55 percent of all calls out of a particular city during the busy time of the day.

Fiber Optics Communications

Fiber optics is a relatively new method of transmitting voice, data, or video signals. It consists of the transmission of light signals over very fine optical fibers bundled together in a flexible cable. It affords extremely high frequencies and high broadband capabilities. Fiber optics transmission was originally marketed for short distances, but it has now become practical for longer distances. The light-emitting diode (LED) was initially the most common light source, but laser beams are now used quite extensively. Special optical modems (RS-232C, etc.) and multiplexers are available for use with fiber optics networks. The real advantages of fiber optics as compared to regular copper wire cable are the high bandwidth, infrequent repeater spacing, low bit error rates, light weight, immunity to electromagnetic interference, and increased security, since tapping into the line is more difficult.

Figure 2-4 shows a diagram of an optical fiber and an example of its use. The optical fiber is a dielectric waveguide that consists of a high-conducting core made of glass fiber enclosed in a low-refractive-index material of glass or plastic called *cladding*. Several of the optical fibers can be bundled into a single cable encased in a plastic cover and strengthened with a core of stronger material. The individual optical fiber is about 0.1 millimeters (mm) in diameter and 0.4 mm when jacketed. A bundled cable of nine optical fibers is about 6 mm in diameter.

The optical signal is subject to bandwidth and amplitude reduction as it travels over miles of the waveguide. This can be compensated for, just as with regular phone lines, by adding repeaters every so many miles as needed. Unlike normal phone line repeaters, however, fiber optics repeaters are also regenerators that create new signals for virtually noise-free transmission. Signal strength is affected by temperature changes and tensile loading. The effects can be compensated for to some extent by packing the individual fibers loosely in the sheathed cable bundle.

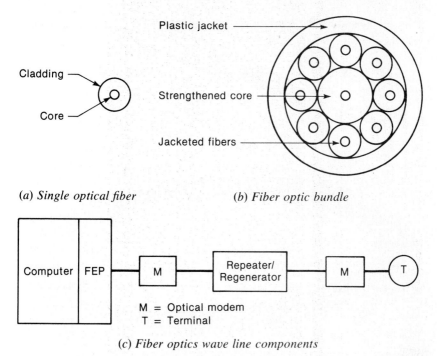

(a) *Single optical fiber*   (b) *Fiber optic bundle*

(c) *Fiber optics wave line components*

Fig. 2-4 Fiber optics lines

Mechanical connectors are employed to splice optical fibers together. If the ends of the fibers are heated so as to melt and become fused, less signal loss and greater reliability can be achieved. The attenuation (signal loss) at an optical fiber splice is about 1.5 decibels (dB) and depends upon correct alignment of the two fibers plus some other factors. Optical fiber cable can be suspended for outdoor runs, but it is usually buried or placed in underground conduit for greater protection.

Coaxial Cables

A coaxial cable is made up of a single wire conductor surrounded by a cylindrical tube of solid or braided mesh conductor. The inner wire and outer tube conductor are separated by either a continuous insulation sleeve or by insulated spacers a fixed number of inches apart (see Fig. 2-5). High-frequency transmission is possible so one common use of

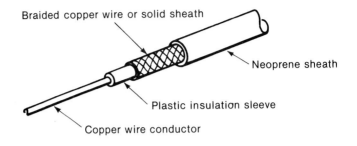

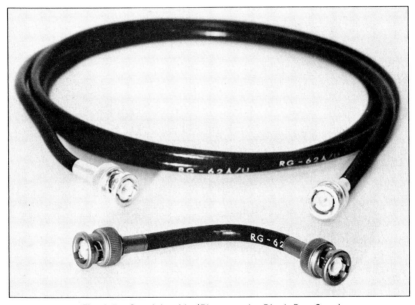

Fig. 2-5 Coaxial cable (Photograph: *Black Box Corp.*)

coaxial cable is for high volume of voice communications and wide-band TV communications. A single coaxial cable can carry up to 4800 time-multiplexed voice conversations employing a 274-Mbps data rate. Coaxial cable is also used for local area networking, to be discussed in Chapter 10.

Smaller, $\frac{1}{4}$-in, coaxial cable has become very popular in data communications work to connect individual CRT displays and printers to cluster controllers. Connectors must be applied to the ends of the coaxial cable to attach the ends to the terminals and cluster controllers or to splice cables together to increase the length of an existing cable run. A special crimping tool is available to apply just the right amount of pressure to secure a copper tip to the center conductor, though this joining can be accomplished with regular pliers if need be.

Coaxial cables are also used for television transmission, as is indicated by the popular term cable television. Though originally installed as one-way broadcast cable systems, television networks can be modified for two-way transmission and can then be used as wideband data and voice transmission media. This prospect was being promoted at the time of writing.

3

Data Communications Line Operational Alternatives

There is a wide array of choices for transmission of data over communications lines. Most of them are made, however, by the terminal or computer manufacturers when they configure their equipment. If the ultimate user does not consider this subject prior to selecting a certain vendor's equipment, the alternatives will be severely restricted. Each user must analyze his or her own particular situation, since what is right for one user may not be right for another. The system application is the key to the correct selection.

The advantages of both leased and dial-up lines were covered in Chapter 2, on Communications Media. To summarize, the rule of thumb is that if there is more than an hour's solid transmission a day, consider the use of a leased line, which is actually the better line of the two. The need for a polling arrangement would dictate the use of a private line. The advantages of analog and digital lines also were covered in Chapter 2. Here a general rule would be to favor a digital line if it is available and suits the particular application being considered within the economic restraints.

Point-to-Point versus Multipoint Networks

A leased line network consisting of one or more computers and multiple remote terminals distributed geographically can be arranged in different configurations depending on such factors as volume of data transmitted and proximity to the host computer and other terminals. The two more obvious alternatives are depicted in Fig. 3-1. It should be noted that the point-to-point network was selected in (a) because each

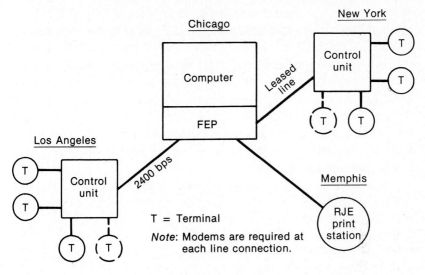

(a) *Point-to-point network*

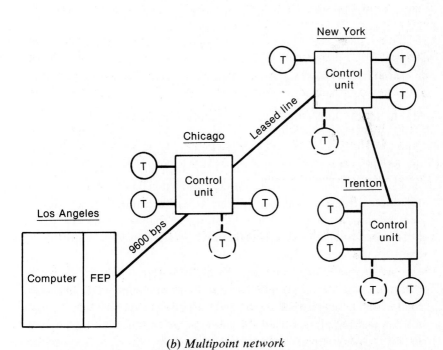

(b) *Multipoint network*

Fig. 3-1 Point-to-point and multipoint networks

of the stations has enough data traffic to justify and fully load a 2400-bps communications line dedicated to only that station. Also, no two stations are on the same route so a higher-speed line could be shared by the two. If this could be made a single multidrop line, it would save the cost of two ports on the front end processor and two modems, but the expensive, long communications lines would have to be shared by all of the using stations. The Los Angeles and New York remote stations will have to be polled, since there are multiple terminals, each with its own address. The RJE print station need not be polled, since it is on a communication line by itself.

A multipoint arrangement was selected in (*b*) because, since the central computer is now in Los Angeles, all three locations are along the same route and can share the same higher-speed communications line. All of the remote locations will have to be polled, since all are on the same communications line and each location has multiple terminals. The normal 3270-type protocol specifies the control unit twice and then the specific terminal twice (e.g., 40-40-C1-C1). Cross-country lines, especially those for public networks, are usually multidrop, and local lines, say, 20 mi long, are usually point-to-point.

Half Duplex versus Full Duplex Transmission

By definition, half duplex transmission means to transmit data over a communications line in only one direction at any one time. This requires the line to be "turned around" constantly to permit the receiving location to send back control codes, such as acknowledgment of a block of good data, or to transmit data when it is that location's turn. In full duplex transmission, on the other hand, data and control codes are sent by, and received by, both locations at the same time. There is a little relief with four-wire half duplex operation in that control codes can be sent back over the extra two wires.

A single transmission path normally requires two wires, so the dial-up network is a two-wire system. Even on your regular telephone the red and green wires are used and the yellow and black wires remain idle or are sometimes used for a second telephone line to the same location. A full duplex transmission line would, therefore, require four wires to obtain the two communications paths. Of course, the second path could be obtained by assigning another frequency and thus running at full duplex on only two wires (e.g., 212A-type modems). However, that is not always practical, especially at high speeds.

Even though full duplex operation is far superior to half duplex in many ways, a survey of current data communications networks would probably reveal half duplex to be still in the majority but steadily losing ground. The reason is that a full duplex operation is normally more expensive; only in recent years have the advantages been recognized as justifying the initial extra expense. The advent of satellite data communications and IBM's SNA concept have promoted the switch to full duplex. Needless to say, the full duplex operation permits far greater throughput of data because not only is transmission taking place in both directions but there is no need to stop and turn the line around to transmit in the reverse direction, which could take 50 ms or more. The cost of the communications line is only about 10 percent more for a four-wire (full duplex) leased line than for a two-wire (half duplex) leased line. It is the terminal or computer that constitutes the greatest part of the cost of full duplex operation.

There is also a simplex communications line by means of which data is sent in only one direction. This type of line is not encountered very often in today's data communications environment. One application is the loop connection arrangement for automated scanning registers in a supermarket system.

Asynchronous versus Synchronous Communications

Asynchronous and synchronous transmission, the two popular modes of data communications, differ in the method of identifying the first and last bits of data that make up a character. Asynchronous transmission is the earlier approach used primarily on simpler terminals like teletypes (Teletype Corp. name for teletypewriter machines). The data bits that comprise a character are identified as being the five or eight bits (Baudot or ASCII codes) that are sandwiched between a start bit and a stop bit, the latter being either $1\frac{1}{2}$ or 2 times the duration of a normal bit in time. Synchronous transmission was developed later, and it is used for the more sophisticated terminals. Here the data bits that make up a character (usually eight for ASCII or EBCDIC codes) are identified strictly by the element of time. This can be initially indicated by one or more *synch characters* emitted by the sending modem, terminal, or front end processor. The synchronous timing character is replenished periodically, as at the beginning of each block of data, to ensure synchronization. Another way to establish synchronization is to use the changes in the status of the bits to check the clocking (i.e., bit synchronization).

Thus we see that, with asynchronous communications, at least $2\frac{1}{2}$ bits are wasted with each character. Asynchronous transmission is still very popular (comprising an estimated 70 percent of all terminals), and it has even found a place in the 3270-type synchronous networks in which a protocol converter permits the less-expensive asynchronous terminal to communicate successfully in the more sophisticated and more expensive synchronous world.

Asynchronous Communications

Asynchronous communications, often referred to as start-stop, is normally restricted to the lower speeds of up to 1200 bps. This term originated from its use with telegraph equipment; a depression of the telegraph key generated a start bit, known as a space bit, with no voltage (or a negative voltage) on the line. A release of the key resulted in a stop bit being generated; it was identified as a positive voltage. The stop positive voltage remained on until the next character started, which resulted in long lapses of time between characters if the operating person was slow. Asynchronous transmission is still used today, but the actual bit pulses of marks and spaces (or binary 1s and 0s as they are referred to today) are generated automatically as complete characters. The bit pulses usually last only a few milliseconds, and the last stop bit of a character is further identified by its time duration being $1\frac{1}{2}$ to 2 times that of the others.

There can still be long periods of inactivity between characters if an asynchronous terminal, such as a teletype machine, is transmitting in the *character mode* and data is being entered manually via a keyboard. There is the alternative of encoding data onto a punched paper tape, magnetic tape cassette, or a diskette while off-line (i.e., not connected to a communications line). Later the terminal can be placed on-line and the data transmitted at the full operating speed of the terminal. This approach could represent a considerable savings in connection time line costs if a dial-up type line is being utilized and is paid for by the minute of usage.

Another variation of asynchronous communications, called *isochronous,* is a kind of mixture of asynchronous and synchronous transmission procedures. The data is framed with start/stop bits as in asynchronous communications, but it is also timed as in synchronous communications to permit higher transmission speeds. Isochronous communications was originally used with the IBM 2260 CRT displays but was later dropped in favor of the 3270 protocol.

Asynchronous terminals have the option of transmission using a *current loop interface* rather than the more common *voltage level interface* defined by the Electronic Industries Association (EIA) standards, like the popular RS-232C. Teletype machines fall in this category. Instead of voltage levels representing the binary 0s and 1s that define bits to make up characters, the current loop interface defines these bits by the absence (0) or presence (1) of a current. The current can be 20 or 60 milliamperes (mA), and a negative current can be substituted for the absence of current. The result, which has been named double current signaling, is more popular in Europe than in the United States.

Synchronous Communications

Synchronous communications permits transmission of blocks of data in a continuous form. Depending on the method used, this mode of data communications can be much more efficient than asynchronous communications. The individual start-stop and even the parity bits associated with each character can be eliminated: synchronization and error detection/correction information can be transmitted only once with each block of data.

Synchronous communications lends itself to the higher transmission speeds of 2400 to 56,000 bps and upward. Transmission schemes for synchronous communications also provide the capability for having multiple information paths or channels via a common communications link. For example, varying the phase of a modem carrier signal in multiples of 90° (degree) increments permits four distinct states per bit time interval, states that can be represented by 00, 01, 10, and 11. Thus two separate streams of data can be transmitted at a 2400-baud rate, which results in an effective information transmission rate of 4800 bps. More detailed information on modem theory is given in Chapter 4.

The synchronization that provides the timing needed to properly locate bits of data can be supplied by bit synchronization, character synchronization, or block (or message) synchronization. In all three methods, the transmitting and receiving terminals employ precise timing circuits, or clocks, which are set to the transmission rate—2400, 4800, 9600, etc., bps. Bit synchronization is accomplished at the receive terminal by sampling the 1-to-0 or 0-to-1 signal transitions of the incoming data stream to set the receive clock to the proper frequency. Since synchronization is lost between line turnarounds, an opening "pad" of alternating 1s and 0s can be transmitted to recapture the synchronization. The popular SDLC protocol (to be discussed later in this Chapter) uses bit synchronization for system timing.

Character synchronization is based on the periodic transmission of special synch characters which are recognized as being nondata characters. If a synch character is not received within a specified time interval, the receiving station will reject all incoming data and look for the synch characters. Synch characters always precede a transmission, and they are usually inserted between blocks of data as well as within the blocks of data. In binary synchronous protocol, character synchronization is used.

Block or message synchronization concerns the framing of data so that blocks of data can be recognized in accordance with the associated data link control procedure.

Code Structures

The basic unit of digital code structures is the bit. It could be described as a low voltage-signal of about 0.02–3 ms in duration (transmission speeds of 300–56,000 bps). During that short period of time, the voltage signal will either be ON to indicate a 1 bit or OFF to indicate a 0 bit. It will take eight of these bits in most code structures to define a single character (e.g., 11000001 is the letter A). Most data transmissions today are performed serially by bit, though there can be an occasional parallel transmission in which all of the bits of a character travel at the same instant at different frequencies over the same wire. Some of the more popular code structures are described below.

Baudot Code

The Baudot code is one of the earliest code structures, and it is still in use today—primarily for low-speed teletypes such as those of the International Telex Network. It has only five bits, which would limit the available character set to only 32, were it not that a system has been devised to double the character set. For instance, a 01010 preceded by a figure shift character represents the number 4; but if the preceeding shift character was a letter shift, the 01010 becomes the letter R.

ASCII Code

The ASCII is probably the most popular code; it is employed in the higher-speed teletypes and is also used for many other data entry terminals. It consists of eight bits per character, and the last bit can be used for parity checking. For example, the letter A is defined by the seven bits 1000001; and since there are an even number of 1 bits, a 1 bit will be added in the eighth position to satisfy an *odd parity* code structure. If one bit (or an odd number of bits) is destroyed during transmission by a line hit or some type of line interference, the character will be detected as being in error. Depending upon the sophistication of the terminal in use at the time, a parity error can result in the retransmission or at least the identification of a bad character.

Table 3-1 shows the structure of the American Standard Code for Information Exchange (ASCII). It should be noted that the bits are received in sequence from right to left 1 to 7 (765 4321) and the bit 8 is the party indicator, odd or even. The equivalent hexadecimal character is determined by adding up the 1 bits which make up the two four-bit halves of a byte. The value of each bit depends upon its position as follows: 421 8421. For instance, 41 = 100 0001, since $4 + 0 + 0 = 4$ and $0 + 0 + 0 + 1 = 1$. This can also be verified by looking at Table 3-2.

Table 3-1 ASCII Code Structure.

Character	Hex	Binary		Character	Hex	Binary	
A	41	100	0001	t	74	111	0100
B	42	100	0010	u	75	111	0101
C	43	100	0011	v	76	111	0110
D	44	100	0100	w	77	111	0111
E	45	100	0101	x	78	111	1000
F	46	100	0110	y	79	111	1001
G	47	100	0111	z	7A	111	1010
H	48	100	1000	0	30	011	0000
I	49	100	1001	1	31	011	0001
J	4A	100	1010	2	32	011	0010
K	4B	100	1011	3	33	011	0011
L	4C	100	1100	4	34	011	0100
M	4D	100	1101	5	35	011	0101
N	4E	100	1110	6	36	011	0110
O	4F	100	1111	7	37	011	0111
P	50	101	0000	8	38	011	1000
Q	51	101	0001	9	39	011	1001
R	52	101	0010	Space	20	010	0000
S	53	101	0011	!	21	010	0001
T	54	101	0100	"	22	010	0010
U	55	101	0101	#	23	010	0011
V	56	101	0110	$	24	010	0100
X	58	101	1000	%	25	010	0101
Y	59	101	1001	&	26	010	0110
Z	5A	101	1010	'	27	010	0111
a	61	110	0001	(	28	010	1000
b	62	110	0010	)	29	010	1001
c	63	110	0011	*	2A	010	1010
d	64	110	0100	+	2B	010	1011
e	65	110	0101	,	2C	010	1100
f	66	110	0110	-	2D	010	1101
g	67	110	0111	.	2E	010	1110
h	68	110	1000	/	2F	010	1111
i	69	110	1001	:	3A	011	1010
j	6A	110	1010	;	3B	011	1011
k	6B	110	1011	<	3C	011	1100
l	6C	110	1100	=	3D	011	1101
m	6D	110	1101	>	3E	011	1110
n	6E	110	1110	?	3F	011	1111
o	6F	110	1111	@	40	100	0000
p	70	111	0000	[	5B	101	1011
q	71	111	0001	\	5C	101	1100
r	72	111	0010	]	5D	101	1101
s	73	111	0011	¬	5E	101	1110

Table 3-1 (*continued*)

Character	Hex	Binary		Character	Hex	Binary	
—	5F	101	1111	ETB	17	001	0111
`	60	110	0000	ETX	03	000	0011
{	7B	111	1011	ACK	06	000	0110
:	7C	111	1100	FF	0C	000	1100
}	7D	111	1101	FS	1C	001	1100
~	7E	111	1110	GS	1D	001	1101
BEL	07	000	0111	HT	09	000	1001
BS	08	000	1000	LF	0A	000	1010
CAN	18	001	1000	NAK	15	001	0101
CR	0D	000	1101	NUL	00	000	0000
DC1	11	001	0001	RS	1E	001	1110
DC2	12	001	0010	SI	0F	000	1111
DC3	13	001	0011	SO	0E	000	1110
DC4	14	001	0100	SOH	01	000	0001
DEL	7F	111	1111	STX	02	000	0010
DLE	10	001	0000	SUB	1A	001	1010
EM	19	001	1001	SYN	16	001	0110
ENQ	05	000	0101	US	1F	001	1111
EOT	04	000	0100	VT	0B	000	1011
ESC	1B	001	1011				

Further, 3C = 011 1100, since $0 + 2 + 1 = 3$ and $8 + 4 + 0 + 0 = 12$, which, according to Table 3-2, represents the letter C in hexadecimal.

Figure 3-2 shows a comparison of the five-level Baudot and the eight-level ASCII codes. In both instances ON and OFF bit pulses are used to identify a particular character, but more bits are required in the ASCII code structure to expand the character set to 128.

(*a*) *The letter R in five-bit Baudot code*

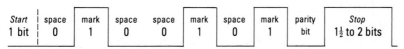

(*b*) *The letter R in eight-bit ASCII code*

Fig. 3-2   Codes for the letter R

Table 3-2 Hexadecimal Numbering Structure.

Character	Half Byte (8-4-2-1)	Sum of Bits
0	0 0 0 0	0
1	0 0 0 1	1
2	0 0 1 0	2
3	0 0 1 1	3
4	0 1 0 0	4
5	0 1 0 1	5
6	0 1 1 0	6
7	0 1 1 1	7
8	1 0 0 0	8
9	1 0 0 1	9
A	1 0 1 0	10
B	1 0 1 1	11
C	1 1 0 0	12
D	1 1 0 1	13
E	1 1 1 0	14
F	1 1 1 1	15

EBCDIC Code

The EBCDIC code was devised by a computer manufacturer as an internal structure to identify characters in memory or storage. When communicating with a computer, a terminal could operate more efficiently if it used the computer's internal code structure. However, because of cost considerations and previous commitments, most terminals transmit only in ASCII, so a code conversion to EBCDIC must be made in the computer's front end processor. The EBCDIC code requires eight bits to identify a character (e.g., A = 1100 0001). A parity bit is not added for communications error detection, the latter being accomplished by a better method to be described later in this chapter.

The eight bits in an EBCDIC character are usually expressed in hexadecimal as two four-bit bytes. The four bits have the values 8, 4, 2, and 1, left to right, and the sum of the byte indicates the hexadecimal character as shown in Table 3-2. We see that a NAK (negative acknowledgment) character is a hexadecimal 3D with the bit structure 0011 (0 + 0 + 2 + 1, or 3) 1101 (8 + 4 + 0 + 1, or D). Table 3-3 shows

Table 3-3 EBCDIC Code Structure.

Character	Hex	Binary		Character	Hex	Binary	
A	C1	1100	0001	e	85	1000	0101
B	C2	1100	0010	f	86	1000	0110
C	C3	1100	0011	g	87	1000	0111
D	C4	1100	0100	h	88	1000	1000
E	C5	1100	0101	i	89	1000	1001
F	C6	1100	0110	j	91	1001	0001
G	C7	1100	0111	k	92	1001	0010
H	C8	1100	1000	l	93	1001	0011
I	C9	1100	1001	m	94	1001	0100
J	D1	1101	0001	n	95	1001	0101
K	D2	1101	0010	o	96	1001	0110
L	D3	1101	0011	p	97	1001	0111
M	D4	1101	0100	q	98	1001	1000
N	D5	1101	0101	r	99	1001	1001
O	D6	1101	0110	s	A2	1010	0010
P	D7	1101	0111	t	A3	1010	0011
Q	D8	1101	1000	u	A4	1010	0100
R	D9	1101	1001	v	A5	1010	0101
S	E2	1110	0010	w	A6	1010	0110
T	E3	1110	0011	x	A7	1010	0111
U	E4	1110	0100	y	A8	1010	1000
V	E5	1110	0101	z	A9	1010	1001
W	E6	1110	0110	0	F0	1111	0000
X	E7	1110	0111	1	F1	1111	0001
Y	E8	1110	1000	2	F2	1111	0010
Z	E9	1110	1001	3	F3	1111	0011
a	81	1000	0001	4	F4	1111	0100
b	82	1000	0010	5	F5	1111	0101
c	83	1000	0011	6	F6	1111	0110
d	84	1000	0100	7	F7	1111	0111

(*continued*)

Table 3-3 (*continued*)

Character	Hex	Binary		Character	Hex	Binary	
8	F8	1111	1000	DC4	3C	0011	1100
9	F9	1111	1001	DEL	07	0000	0111
&	50	0101	0000	DLE	10	0001	0000
-	60	0110	0000	DS	20	0010	0000
/	61	0110	0001	EM	19	0001	1001
$	5B	0101	1011	ENQ	2D	0010	1101
¢	4A	0100	1010	†EOB	26	0010	0110
!	5A	0101	1010	EOT	37	0011	0111
:	7A	0111	1010	‡ESC	27	0010	0111
#	7B	0110	1011	†ETB	26	0010	0110
,	6B	0110	1011	ETX	03	0000	0011
.	4B	0100	1011	FF	0C	0000	1100
<	4C	0100	1100	FS	22	0010	0010
*	5C	0101	1100	HT	05	0000	0101
%	6C	0110	1100	IFS	1C	0001	1100
@	7C	0111	1100	IGS	1D	0001	1101
(	4D	0100	1101	IL	17	0001	0111
)	5D	0101	1101	IRS	1E	0001	1110
–	6D	0110	1101	IUS	1F	0001	1111
'	7D	0111	1101	LC	06	0000	0110
+	4E	0100	1110	LF	25	0010	0101
;	5E	0101	1110	NAK	3D	0011	1101
>	6E	0110	1110	NL	15	0001	0101
=	7E	0111	1110	NUL	00	0000	0000
¦	4F	0100	1111	PF	04	0000	0100
—ˆ	5F	0101	1111	PN	34	0011	0100
?	6F	0110	1111	‡PRE	27	0010	0111
"	7F	0111	1111	RES	14	0001	0100
{	C0	1100	0000	RLF	09	0000	1001
}	D0	1101	0000	ACK	2E	0010	1110
\	E0	1110	0000	RS	35	0011	0101
~	A1	1010	0001	SI	0F	0000	1111
	79	0111	1001	SM	2A	0010	1010
	6A	0110	1010	SMM	0A	0000	1010
BEL	2F	0010	1111	SO	0E	0000	1110
BS	16	0001	0110	SOH	01	0000	0001
BYP	24	0010	0100	SOS	21	0010	0001
CAN	18	0001	1000	Space	40	0100	0000
CC	1A	0001	1010	STX	02	0000	0010
CR	0D	0000	1101	SUB	3F	0011	1111
DC1	11	0001	0001	SYN	32	0011	0010
DC2	12	0001	0010	UC	36	0011	0110
DC3	13	0001	0011	VT	0B	0000	1011

† EOB and ETB have the same hex assignment.
‡ ESC and PRE have the same hex assignment.

the code structure for the Extended Binary Coded Decimal Interchange Code (EBCDIC). Note that 256 characters are available, as compared to only 128 for the ASCII code structure. The bits are received in sequence from right to left when looking at the hexadecimal bit structure of 8765 4321.

Four-of-Eight Code

The four-of-eight code was devised for error detection: four of the eight bits transmitted must always be 0s and four must be 1s. This gets around the vulnerability of the ASCII code error detection scheme consisting of a single parity bit, which could fail if two bits were to be changed by a transmission line error. The four-of-eight code is not in very popular use today.

Communications Traffic Protocols

Protocol Level Definitions

A set of rules is required to regulate the data communications traffic between two or more locations in a network. Some of the possible problems are two locations transmitting at the same time on a half duplex line, one location transmitting when the receiving location is not ready, and the retransmission of erroneous data.

The International Standards Organization (ISO) has identified the different levels of protocol that are required to coordinate all of the functions needed to permit terminals and computers to communicate with each other in a modern data communications network. This modular design of a communications network isolates the various communications functions into standardized layers which are independent of each other. Thus a modification in one of the layers does not require a change in any of the other, independent layers. The seven levels of protocol are defined below.

Level 1 covers the physical and electrical interface between the terminal and the network. This has been further identified by the EIA and CCITT standards dealing with such items as voltage levels, data interchange, control circuits, impedance, and speed. The EIA RS-232C, for example, specifies those standards.

Level 2 is concerned with data link control (DLC), also referred to as communications line discipline. It is the set of rules for operation between computers and terminals that results in an orderly transmission of blocks or frames of data. The DLC protocol is the best known level; we usually mean it when we use the term protocol in reference to start-stop, bisynchronous, SDLC, etc. An example of the sequence of events that take place when a computer polls a terminal as part of the level 1 and 2 protocols is described below. The reference is to the EIA cable pin definitions (RS-232C) shown in Appendix B.

- The host computer raises REQUEST TO SEND.
- The host modem sends a MARK frequency to the remote modem.
- The remote modem raises CARRIER DETECTOR and notifies the remote terminal.
- The host modem raises CLEAR TO SEND.
- The host computer sends the poll data to the host modem on the TRANSMIT DATA pin circuit which, in turn, sends it to the remote modem.
- The remote modem sends the poll data to the remote terminal on the RECEIVE DATA pin for decoding.
- The remote terminal raises REQUEST TO SEND, and a MARK frequency is transmitted from the remote modem to the host modem.
- The host modem raises CARRIER DETECTOR and notifies the host computer.
- The remote modem raises CLEAR TO SEND.
- The remote terminal sends a poll response to the remote modem on the TRANSMIT DATA pin.
- The remote modem sends the poll response out on the communications line to the host modem, which, in turn, sends it to the host computer on the RECEIVE DATA pin.

Level 3, the "network level," deals with the control procedures and format required to move data around the nodes of a large communications network. It differs from levels 1 and 2 in that it is accomplished entirely by software rather than hardware. Level 3 protocol is used to handle several streams of data across the same link by individual addressing. This is more prominent in the larger, public networks, where multiple users share the same communications lines.

Level 4 is concerned with the assembly of messages from blocks of data and the problems associated with loss or duplication of data. It is called the *transport level*.

Level 5 deals with initialization, running, and termination of sessions. It is called the *session level*.

Level 6 covers the interpretation of exchanged data, display of data, character codes, data structure, and formats. It is called the *presentation level*.

Level 7 deals with the actual application of such functions as file access, batch processing control, and file transfer protocol. It is called the *application level*.

Asynchronous Protocol

The asynchronous protocol is very simple, but it has served its purpose for a long time. The start bit triggers the receiving station to start sampling data bits, which are the next five or eight signal pulses. Then the longer stop bit identifies the end of the character. There can be a lengthy pause between characters if the operator's terminal does not have a buffer, paper tape, cassette, etc., to provide a steady flow of characters.

Asynchronous transmission usually consists of a string of characters of undetermined length. In many cases, if the communications connection is disrupted, the sender continues to transmit even though the receiving station is not receiving the data. Most receiving terminals have some error detection ability (parity check on ASCII code or invalid bit arrangement of Baudot code), whereby invalid characters are identified with an E or a SUB code. Some of the more sophisticated terminals can request a retransmission of previously transmitted data containing errors.

Another form of error detection and correction that could be used is the *echo method*. The sender transmits one character and stops. The receiver sends it back; and if it is different (usually on a CRT display), it will be transmitted again. The price one pays here is an over 50 percent reduction in line speed.

Binary Synchronous Protocol

The binary synchronous protocol is most popular at this time. It will accommodate the ASCII, EBCDIC, or even the lesser used Six Bit Transcode. The code set is comprised of data characters (e.g., 0 = 9, A, B, C, etc.), device operational characters (e.g., horizontal tab), and data link control (DLC) characters (e.g., start of header). The latter are

Table 3-4   **Binary Synchronous Data Link Control Characters.**

Code	Hex	Explanation
SYN	32	Synchronous idle—establish timing synchronization.
ENQ	2D	Enquiry—Bid for line or repeat of a transmission.
DLE	10	Data Link Escape—Precedes supplementary line control characters and transparent mode control characters.
SOH	01	Start of Header—Optional header information.
STX	02	Start of Text—Identifies the data.
ITB	1F	End Intermediate Block—Sends block check character but avoids line turnaround.
ETB	26	End of Transmission Block—Block check character and line turnaround for ACK or NAK.
ACK0 or ACK1	1070 1061	Affirmative Acknowledgment—Alternate 0 and 1 recognizes a loss of acceptance code.
NAK	3D	Negative Acknowledgment—Calls for retransmission.
EXT	03	End of Text—Identifies end of data.
EOT	37	End of Transmission—No more to send. Also, negative response to a poll or indicates an abort.
WACK	106B	Wait before Transmit—Positive acknowledgment plus receiving station not ready to receive more.
RVI	107C	Reverse Interrupt—Positive acknowledgment plus receiving station urgently wants to transmit.
TTD	022D	Temporary Text Delay—Sender can't transmit but wants to hold line.
BCC	Varies	Block Check Character—Final bit count in hexadecimal.
DLE/ EOT	1037	Disconnect for Dial-Up Line.
BEGIN PAD	55	Beginning Pad—Filler to protect first character.
END PAD	FF	Ending Pad—Filler to protect last character.

shown in Table 3-4. There is also the option to transmit in the *transparent mode* for data such as computer programs so characters that look like data link control characters will not be acted upon as such. The beginning and ending of text data must be preceded by DLE characters as well as any control characters that need to be recognized.

ERROR DETECTION AND CORRECTION

The error detection and correction procedures provided for binary synchronous communications are far better than those provided for asynchronous communications. They can check the parity of each character if the ASCII code is used, and they can also perform either vertical or cyclical redundancy checking of any code employed. Either a summation (vertical) or a mathematical calculation (cyclical) is performed by using all of the bits at the transmitting station. The resulting count is sent at the end of a block of data (BCC character). The receiving station then recalculates the *block count* and compares that count with the sender's count. A disagreement results in a negative response to the sender, who must then retransmit the entire block of data. There is an advantage to transmitting fairly large blocks of data, say, over 200 characters. This is particularly true in half duplex operation, in which time-consuming line turnarounds can be decreased. This could also become a disadvantage if there are many line errors, since the entire block of data must be retransmitted.

LINE CONTROL

The general sequence of line control characters is SYN-SYN-SOH (optional)-STX-Text-ITB or ETB- or ETX-BCC-EOT. However, the exact use of the protocol characters depends upon the devices at each end and their communications relations. One very common relation is called *point-to-point with contention.* Here there are only two terminals, or computers, on the line and each must contend for the privilege of transmitting its data. The communications line can be either leased or dial-up. The control character sequence is shown in Table 3-5. Another common relation is the *multipoint operation with polling* from a central location. The control character sequence is shown in Table 3-6. Here the *general poll* approach is used; it is usually preferred as being more efficient. The remote control unit interrogates each of its up to 32 terminals (CRT displays and printers) to see if any have data or status or test messages to transmit.

Table 3-5   Example of Point-to-Point Bisynch Protocol.

Point A Point B	Explanation
SYN ———————→ SYN ENQ	Point A bids for the line.
←——————— SYN SYN ACK0	Point B gives A the line.
SYN ———————→ SYN SOH Header STX Text ETX BCC	Point A sends header and text data.
←——————— SYN SYN ACK1	Point B recalculates block check character and then gives a positive acknowledgment.
SYN ———————→ SYN EOT	Point A ends the transmission.

Table 3-6   Example of Bisynch, Multipoint, General Polling.

Computer Center Remote CU	Explanation
PAD ———————→ SYN SYN EOT C3 C3 7F 7F ENQ PAD	The computer's front end processor sends out a general poll (7F) to remote cluster control unit number 4 (C3).
←——————— PAD SYN SYN EOT PAD	The remote control unit has no data to transmit from any of its displays, so it responds with an EOT.

A *specific poll* approach also can be used: a specific terminal address replaces the general poll's 7F7F address. For example, C3 C3 40 40 would interrogate only the first terminal port (40) on the fourth control unit (C3) for data to send. It would not acknowledge requests by other terminal ports that may have data to send at that particular time. Table 3-7 shows a specific poll for the tenth terminal (4A) on the first control unit (40), and this time there is data to transmit.

A third approach would be to employ "selection addressing," which initiates an immediate transfer of the buffered data from the selected terminal to the remote control unit. The addressing would be something like 60 60 C1 C1 for immediate selection of terminal 2 (C1) on control unit 1 (60).

Table 3-7   Example of Bisynch, Multipoint, Specific Polling.

Computer Center	Remote CU	Explanation
PAD ⟶ SYN SYN EOT 40 40 4A 4A ENQ PAD		The computer sends out a specific poll.
⟵	PAD SYN SYN SOH Header STX Text ETB BCC PAD	The terminal has data and sends it back.
PAD ⟶ SYN SYN ACK0 PAD		The computer acknowledges a correct receipt with an ACK0.

SDLC Protocol

The *S*ynchronous *D*ata *L*ink *C*ontrol (SDLC) protocol is a more recent innovation that is gaining in popularity, along with some other bit-oriented protocols. Here bit synchronization must be provided by either the modem or the data terminal equipment (terminal, control unit, or front end processor). It can be operated as half or full duplex, or the primary station (computer) can be operated at full duplex and the secondary stations at half duplex. Since the data block is the basic record for bisynchronous protocol, the comparable "frame" is the basic record for all SDLC information exchange between locations. The frame is shown in Fig. 3-3, and the fields are described below.

Beginning Frame Flag—Always 01111110. It identifies the beginning of a new frame. Any transmitting station is required to add a zero after any sixth consecutive 1 bit it may be sending as data, and then the receiving station will remove the 0. This prevents any data from being mistaken for a flag field.

Address Field—Address of receiving location, eight bits.

Control Field—In the *information transfer format* there are three bits for frames-sent count, one bit for a poll request or response indicator, three bits for frames-received count, and one extra bit. This is a means of verifying frames sent; it triggers a retransmission if frames are lost and also identifies any duplicates. In the *supervisory and nonsequenced formats,* the control field can contain commands and responses such as reject, test, and disconnect.

Information Field—Contains the actual data to be sent. It is variable in length in multiples of eight bits.

Frame Check Sequence Field—This 16-bit field contains a cyclical redundancy checking number (same as for bisynch protocol). Any frames in which the sender's and receiver's counts do not agree must be retransmitted. Seven frames can be sent before an

Flag	Address	Control	Information	Frame check sequence	Flag

Fig. 3-3   SDLC frame structure

acknowledgment must be sent back, and the acknowledgment can be sent in the same frame as any data available. There are SDLC procedures, however, that permit the option to avoid a line turnaround after every seven frames. Frames being transmitted can specify no response, response on exception only, or a definite response. Frames can also be *chained* for long uninterrupted transmissions or even *bracketed* for even longer transmissions. Another bit-oriented protocol, DECNET, provides for 255 outstanding frames, and HDLC provides for 128.

Ending Flag—Defines end of frame as 01111110.

The command and response codes used under SDLC are described in Table 3-8. Note that there are often two hexadecimal designations for each code, one with the poll or final bit ON (P or F) and one with it OFF ($\overline{P}$ or $\overline{F}$). The poll bit ON demands a transmission from the secondary station. The final bit ON indicates the end of a transmission. Ns = frames sequence numbered by sending station. Nr = error-free frames sequence numbered by receiving station.

Table 3-9 describes the operation of a multipoint full duplex operation under SDLC.

The SDLC procedure also permits the use of a *loop network* that operates under half duplex only. Each secondary station acts as a repeater to relay signals initiated by the single primary station. Of course, if a secondary station recognizes its own address, it captures the frame for its own use. The primary station can initiate a polling request by transmitting continuous 1s. The next station down the loop can transmit any frames of data it desires and go back into the repeater state so the next station can then transmit if it so desires.

X.25 Protocol

The X.25 protocol is a recently established bit-oriented protocol that has been accepted quite universally in the United States and Europe. It is used extensively in public packet switching networks, whereby data transmitted by member companies is broken into small packets of perhaps 128 characters and intermixed (multiplexed) on the same high-speed lines, later to be reassembled into intelligible messages at the destination. The transmissions are point-to-point, synchronous, and usually in ASCII code for packet switching networks.

Table 3-8 SDLC Command and Response Codes.

Code	Hex (ON)	Hex (OFF)	Explanation
SNRM	93	83	**Nonsequenced Commands** Set Normal Response Mode—Receiving station is set as a secondary, subordinate station.
DISC	53	43	Disconnect—Receiving station is made inactive and placed off line if a leased line or "on hook" if a dial-up line.
SIM	17	07	Set Initialization Mode—Receiving station is prepared for forthcoming link level functions.
NSI	13	03	Nonsequenced Information—Identifies nonsequenced information format (for commands, etc.).
NSP	33	27	Nonsequenced Poll—Receiving station is invited to transmit. Usually used for loop applications.
XID	BF		Exchange Station Identification—Receiving station is requested to send identification. Usually used for dial-up stations.
TEST	F3	E3	Test—Receiving station is requested to send back any information field received as a test.
NSA	73	63	**Nonsequenced Responses** (also NSI, XID, and TEST) Nonsequenced Acknowledgment—Affirmative response to SNRM, DISK and SIM commands.
ROL	1F	0F	Request On-line—Sent by secondary station to indicate that it will be disconnected.
CMDR	97	87	Command Reject—Sent by secondary station to indicate nonvalid command received.
RQI	17	07	Request for Information—Sent by secondary station to request an SIM command.
RR	†1		**Supervisory Commands and Responses** Receive Ready—Either station is ready to receive and confirms sequenced frames 0 through 7 via the * position digit.
RNR	†5		Receive Not Ready—Either station indicates a busy condition, confirms last sequenced frame received, and requests next frame via the * position digit.
REJ	†9		Reject—Either station requests retransmission of frames 0 through 7 via the * position digit.
I	**		**Information Commands and Responses** Information—Either station sends an information formatted frame. The first * indicates the count of frames 0 through 7 and the poll or final bit ON or OFF. The second * indicates the count of frames sent.

† The characters in these positions vary from 0 to E according to a special cross-reference chart that codes the various combinations of Nr, Ns, P, P̄, F, and F̄.

Table 3-9   Example of SDLC, Multipoint with Polling.

Computer Center	Remote CU-D	Remote CU-E	Explanation
D ⟶ SNRM P			Computer center sets station D on-line and resets frame checking sequence field by using set normal response mode command code. D = remote D's address. P = poll bit is set ON.
⟵	D NSA F		Remote D acknowledges with a non-sequenced acknowledgment code. F = final frame bit ON.
E ⟶ RR P(0)			Computer center polls remote E for a transmission by using a receive ready command. P = poll bit is set ON. (0) = set first frame count.
D I(0) P̄(0)			Computer center sends a frame to remote D using information transfer format. (0) = set first send frame count. P̄ = poll bit is set OFF.
⟵		E I(0) F̄(0)	(0) = set first receive frame count. At the same time, remote E sends frame to computer center by using information format. (0) = set first send frame count. F = final frame bit is set ON. (0) = set first receive frame count.
D ⟶ RR P(0)			Computer center polls remote D for confirmation by using receive ready command. P = poll bit is set ON. (0) = set first receive frame count.
⟵	D RR F(1)		Remote D acknowledges frame 0 using receive ready command. F = final frame bit is set ON. (1) = frame 0 confirmed received.
E ⟶ RR P̄(1)			Computer center acknowledges frame 0 from E by using receive ready command. P̄ = poll bit is OFF. (1) = frame 0 confirmed received.

X.25 is very similar to SDLC and HDLC, another bit-oriented protocol available, as seen in Fig. 3-4. The most noticeable difference is the addition of a 24-bit PACKET HEADER containing LOGICAL CHANNEL GROUP NO. (4 bits), GENERAL FORMAT IDENTI-FIER (4), LOGICAL CHANNEL (8), and PACKET TYPE IDENTI-FIER (8). In this country, X.25 utilizes the ADCCP (advanced data communications control procedure) for its frame structure.

Microcomputer and Packet Network Protocol

What the users are looking for is an error-free local connection for the "last mile" leading to the user's terminal. Microcomputers are capable of sending and receiving binary files, which are more vulnera-ble to errors than straight text files, so the error rate becomes an important consideration. The users need a protocol that will handle the transmission of binary files more efficiently than the normal X.25 packet switching protocol.

The selection of a communications protocol for microcomputers (including personal computers) that deal with packet switching net-works is open to the choice of the user because no standard has been established as yet. The only standard under consideration at the time of writing was the X.32, which is a synchronous protocol based on an X.25 link to a packet network to a dial-up line being considered by the CCITT. One of the problems with this approach is the increased cost of transmitting synchronously, because many of the microcomputer man-ufacturers are set for the less-expensive asynchronous mode of com-munication. This protocol also deals only with the second (data link) and third (network) levels of the ISO protocol recommendations.

Two other microcomputer packet network protocols are in use today by private installations, and both of them operate asynchro-nously. The MNP protocol was introduced by Microcom and is being offered by both the Telenet and Uninet packet switching networks. The system is based on the use of Microcom's modems, but it covers the greatest number of ISO protocol layers (2, data link; 5, session; 6,

Flag (8)	Frame header (16)	Packet header (24)	Text (up to 1024)	BCC (16)	Flag (8)

Fig. 3-4 X.25 frame structure

presentation; and 7, application). The MNP system is more versatile, since it can be used with almost any microcomputer and need not be confined to a packet switching network.

The X.PC protocol was designed by Tymnet specifically for relatively high performance microcomputers for use on a packet switching network. X.PC deals with fixed header lengths and only one of the ISO protocol layers (2, data link).

Data Encryption

Data encryption is concerned with the manipulation of information to prevent its use by persons other than those for whom it was originally intended and who have been given the key for its proper decoding and use. As the country moves toward increased use of computers and the transmission of data, the need to prevent information from falling into the wrong hands has increased considerably. The expansion of distributed data processing and the use of microwaves and satellites for data communications, in which a physical connection is not required, have made data encryption very important to some of our highly competitive businesses, not to mention the military. All transmissions of data in the banking business require data encryption.

Data encryption is accomplished by a manipulation of the bit patterns which is transparent to the users of the system. The user enters a secret key to encrypt a message, and only the intended receiver has the proper decrypt key to decipher the message or data. There are complex algorithms to substitute and transpose information to make it unusable to anyone who does not possess the key to the decryption. Also, keys are changed as frequently as deemed necessary to make it even more difficult for unauthorized outsiders to gain access to confidential information.

The need for system timing or synchronization must be considered for any encryption requirement. For example, asynchronous communications lines cannot be compatible with some crypto systems or a mix of asynchronous and synchronous terminals can pose problems in system timing. Furthermore, dial-up lines can require one form of security and leased lines can require a lower level of security. A complete systems analysis is recommended prior to any decision to implement encryption in the data communications network.

4

Modems

In appearance, most modems come in a metal or plastic box about the size of an unabridged dictionary. When a quantity of the same modem is needed, there is usually the option of installing a single cabinet with a single power supply that will accommodate perhaps up to 24 modem boards. This choice will save space and a little money, but a backup power supply might be worthwhile. Equipment vendors, and especially of such terminals as teletypes, can build modems right into their equipment cabinets (see Fig. 4-1).

The modem (acronym for *mo*dulator-*dem*odulator) is used to interface data terminal equipment (terminals, computers, printers, etc.) with analog telephone lines. The analog lines cannot otherwise pass high-speed digital data without severe distortion effects. Furthermore, the bandpass of an analog telephone line can, by the employment of multiple tones and modulation techniques, handle a number of data channels simultaneously. For example, the Bell 212A Modem can handle full-duplex operation at data rates up to 1200 bps over a two-wire switched circuit.

Modems can be divided into four basic categories in regard to speed of operation:

- Narrow-band modems—operating up to about 300 bps, are used for low-speed terminals such as teletypewriter machines. These modems utilize the lower end of the voice-grade channels.
- Voice-grade modems—typically operate at speeds of 1200, 2400, 4800, 9600 bps, and above. Special high-speed modems, capable of operation up to 14,400 bps also are available.
- Wideband modems—require a single-channel bandwidth equivalent to 12 voice-grade channels, typically operate at data rates of 19.2 to 64 kbps (or 19,200 to 64,000 bps).

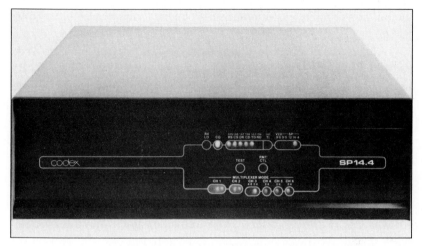

Fig. 4-1 Stand-alone modem (*Codex Corp.*)

- Limited-distance or short-haul modems—can handle data rates up to 1.5 Mbps (or 1,500,000 bps).

Other classifications of modems include asynchronous or synchronous operation, "smart modems" employing microprocessors, voice/data capability, and many other optional features. Some modems even incorporate antiintrusion capability to protect against illegal entry to the data communications system.

Modem Functions

The primary function of a modem is to covert the digital signals from data terminal equipment (computers, CRT terminals, etc.) to a modulated analog carrier signal suitable for transmission over available voice telephone lines. These modulated carrier signals or tones are transmitted at predetermined frequencies within the bandpass of the voice channel. The modem modulates the ON and OFF digital computer or terminal data bits onto a flat carrier signal for transmission over the available voice telephone network. The carrier can be sent at a variety of fixed frequencies, but it is only the means of transportation for the modulated signal, which contains all of the data intelligence. The three modulation techniques—amplitude, frequency, and phase—are shown in Fig. 4-2.

Amplitude modulation is accomplished by varying the amplitude or voltage level of the carrier signal to depict a binary 0 or 1. By increasing the number of amplitude levels, more binary combinations are made possible for transmitting more data in the same time frame. Frequency modulation, often referred to as frequency shift keying (FSK), is accomplished by varying the frequency of the carrier signal. For example, a carrier frequency of 1700 Hz can be shifted +500 Hz (to 2200 Hz) to represent binary 0 and −500 Hz (to 1200 Hz) to represent binary 1. The Bell System Type 202 Modem uses this scheme for half duplex operation at data rates up to 1200 bps with extended operation to 1800 bps. Figure 4-3 shows the frequency graph for this modem. Note that a narrow reverse channel, centered at 387 Hz, is used for positive acknowledgment (ACK) or negative acknowledgment (NAK) of the main data channel transmission.

Phase modulation, also referred to as phase shift keying (PSK), is accomplished by varying the phase of the carrier signal by a predetermined number of electrical degrees. For example, shifting the carrier

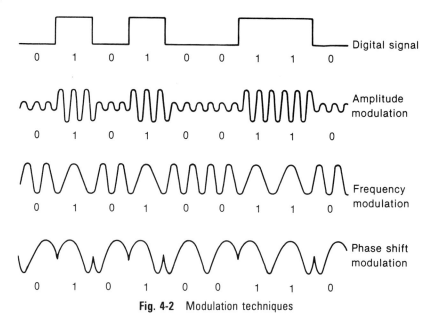

Fig. 4-2 Modulation techniques

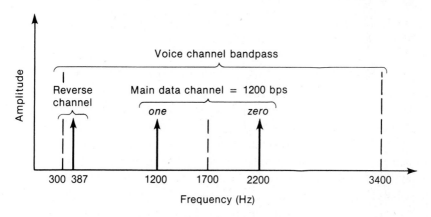

Note: The binary *one* and *zero* signals are transmitted
at 1200 and 2200 Hz, respectively.

Fig. 4-3 Frequency characteristics of the Bell System 202 Modem

180° can be used to indicate a change from binary 1 to binary 0. By increasing the shifting levels to 90 and 270°, the four resulting states of 90, 180, 270, and 360° can be used to represent the binary combinations 00, 01, 10, and 11. This permits a higher data transmission rate within the same time frame. Some modems use phase shifts of only 45° to obtain eight different combinations in multiples of 45°. This results in a threefold increase of data transmission rate as compared to a single channel using 180° phase shifts.

Quadrature amplitude modulation (QAM), in which a combination of phase shift and amplitude modulation techniques is employed, is used to increase channel capacity to 9600 bps and higher. For example, in the Bell 209 and CCITT V.29 standards QAM modulation is used to achieve full duplex, four-wire operation at 9600 bps.

The telephone company has initiated a requirement that is intended to protect switched network users from any equipment that can interfere with the proper operation of the network. A protective device called a data access arrangement (DAA) is required to be placed between the non-Bell modem and the dial-up communications line. Some modems are manufactured with the protective circuitry built into the devices themselves as an added feature. If so, the user must obtain the equipment approval number and ring equivalence before the phone company will provide the dial-up line. Otherwise, a telephone-company-manufactured DAA must be ordered.

INTERFACE BETWEEN MODEM AND BUSINESS MACHINE

Another important function of the modem is to provide the electronic logic boards that regulate the operational interface between the modem (representing the communications line) and the business machine, be it a terminal, cluster control unit, or front end processor. The cable connecting the modem and terminal contains multiple wires such as the 25-pin arrangement of the U.S. standard RS-232C most commonly used. Low-voltage signals are applied to the pins to indicate such communications traffic events as DATA SET READY, DATA TERMINAL READY, REQUEST TO SEND, CLEAR TO SEND, DATA CARRIER DETECTOR, DATA MODULATION DETECTOR, SPEED SELECTOR, TRANSMIT TIMING, RECEIVE TIMING, GROUNDS, TRANSMIT DATA, and RECEIVE DATA.

The timing for a data transmission is usually originated by the modem, but it can come from the front end processor or the terminal control unit if so specified by the manufacturers involved. Some modems are speed-selectable via a switch (e.g., 2400, 4800, or 9600 bps) or can even detect the speed of an incoming signal and accept it.

Most modems have some indicator lights that aid in troubleshooting a line, such as showing that the power is on, that a carrier signal is present so the line is not open, or that the proper interface events (e.g., REQUEST TO SEND) have taken place. Modems also have switches to assist in problem solving, such as "busy out" the modem, self-testing, and "loopback testing" to send a signal to the other end and back again to ensure the modems and line are operational. Certain manufacturers of communications network control systems include additional logic boards that permit interrogation of the modem from the network control center to detect failures or even pending failures.

LINE EQUALIZATION

Another important function of the modem is that of *line equalization*. Analog telephone lines present the problem of attenuation and delay distortion being different at different frequencies. Attenuation concerns the weakening of the signal strength, and delay distortion causes an undesirable echo. By tuning, or equalizing, the modem at each end of the line, the difference can be compensated for and the problem eliminated. The equalization tuning can be done manually at each end with a screwdriver on the less-expensive modems. This tuning is necessary infrequently; but if the telephone company varies the line parameters or something else happens to disturb the balance of the line, the equalization procedure must be repeated. (The phone company's conditioning of a leased line can also help get around the equalization problem.)

RECEIVE DATA TIMING

Some modems provide *receive data timing* to the associated terminal equipment for use with bit synchronization protocols such as SDLC. A receiving station normally samples the incoming signal at the same rate as the signal is transmitted from the sending station. However, variations do occur, and the receiving modem should be able to dynamically adjust the sample timing accordingly. By continual adjustment of the timing alignment, the receiver timing signal is forced into synchronization with the incoming signal and thus the transmitter. If the modem cannot provide this function, the terminal must maintain bit synchronization by employing a technique like the non-return-to-zero inverted coding method (NRZI) in which the signal remains in the same state to send a binary 1 and changes states to indicate a binary 0. This induces transitions in those occurrences of long periods of binary 0s so common to data transmissions. Figure 4-4 shows the RZ, NRZ, and

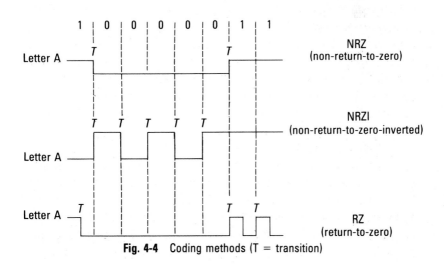

Fig. 4-4 Coding methods (T = transition)

NRZI coding methods. The NRZ code, which was added just for comparison, does not provide transition clocking for either the binary 1 or 0 signals and thus is not very practical for synchronous data transmissions that are bit-oriented.

RECEIVE CARRIER DETECTION

Another modem function is called *receive carrier detection.* The modem contains a circuit that monitors the level of the signal being received by comparing it with a given acceptable level. If the incoming signal level drops below the acceptable level by as much as 1 dB, then after 8 ms the CARRIER DETECT voltage signal is removed. This fixes RECEIVE DATA at a MARK condition and prevents the receipt of any erroneous incoming data.

Modem Features

Most modems are designed for specific applications such as 9600-bps, leased-line, four-wire. Special line conditioning is often needed to run at the higher speeds, like 9600 bps. Multipurpose modems, such as the popular 212A/103A variety that operate at either full duplex at 1200 bps or at half duplex at 300 bps, are the exception. Incidentally, the 212A-type modem is itself a little unusual in that it operates at full duplex by using only two wires. It does so by a form of frequency-division multiplexing that virtually splits the two wires into four wires.

Many modems in the synchronous medium-speed class provide the capability of frequency-division multiplexing to divide a 9600-bps line into two 4800-bps lines or four 2400-bps lines. The perhaps four data stations that will use the four resulting 2400-bps communications channels are normally situated in the same location within about 50 ft of the multiplexing modem. A data station any number of miles away can use one of the multiplexed channels if it is connected as a *tail circuit* as shown in Fig. 4-5. The less-expensive modem eliminator can be used in place of the two modems if the tail circuit is less than about 100 ft away. If the distance is a few miles or so, a limited-distance modem might be best.

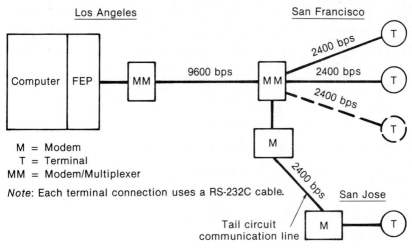

Fig. 4-5   Modem multiplexer with tail circuit

DIAL BACKUP

Some leased line modem manufacturers offer a dial backup feature that can be used in an emergency when the leased line is inoperable because of a line problem. The procedure for going to dial backup usually consists of flipping a switch on the modem at each end and placing two long-distance phone calls. At a small additional expense the remote location can be arranged to answer automatically without operator intervention. Figure 4-6 shows the connection setup.

AUTOMATIC EQUALIZATION

Automatic equalization is a feature that would be important at the higher transmission speeds of 2400 bps and above, where the equalization is needed to compensate for the delay distortion which is often present at the higher speeds. With some less-expensive modems, particularly limited-distance modems (to be discussed later in this Chapter), the person installing the modem has to adjust the equalizing circuit with a screwdriver to set it properly for that particular circuit's conditions. This takes time, and it can have to be done over again if the telephone company ever changes the line conditions. It also adds an extra step to communications line maintenance procedures, since a problem with the line would probably call for a reequalization check.

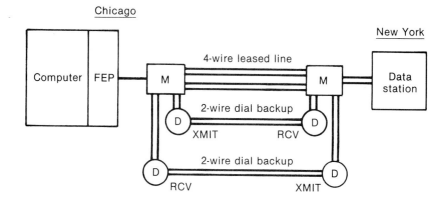

Fig. 4-6 Leased line dial backup arrangement

With automatic equalization the adjustment to compensate for delay distortion is performed automatically by the modem logic at installation time and whenever the condition of the line can be changed.

LINE TURNAROUND

Line turnaround is the time it takes to change the direction of transmission on a half duplex communications line. This can add up to considerable difference, especially when operation is in a conversational mode or when short blocks of data require frequent acknowledgments. It can be important, if not just an advantage, to reduce the line turnaround time to be as low as possible, especially when there is a variation as from 250 ms for some modems and 8 ms for others.

SECONDARY CHANNEL

Some modems provide the option to utilize a secondary channel. This could be an added benefit in some applications such as the installation of a second terminal, e.g., a teletype machine, at each location on a 2400-bps CRT display network. Another popular use of the secondary channel is for a network modem monitoring system from the host computer central location.

LOOPBACK TESTING

Remote and local loopback testing features are of considerable value during problem-resolving operations on a network. The local loopback feature sends a signal created by the modem's test circuits through the modem and back out to determine if the local modem is operating satisfactorily. This test can be performed at each end of a communications line to check the individual modems for malfunctions. A more thorough test would be to perform a remote loopback check in which a signal is generated by the local modem, sent down the communications channel to the remote modem, and circulated back to its source with display lights to indicate the results of the test. In some cases a person is required at the other end to flip a switch to activate the turnaround circuit, and in other cases the circuit can be activated automatically by a special-frequency signal from the originating modem. The remote loopback test has the additional advantage of checking the communications line as well.

ALTERNATE VOICE TRANSMISSION

Another modem feature that can be of value concerns the use of the communications line for alternate voice and data transmission as desired. For instance, a leased line that is used for voice transmissions during the prime time of the workday can be used for data transmissions all night by merely flipping a switch on the modem at each end of the line.

AUTOMATIC ANSWERING

Automatic answering (unattended) for a switched or dial-up line is almost mandatory these days as a modem feature. However, there still can be situations in which it is necessary for a person at a data center or a remote terminal location to answer the telephone and manually switch to the data transmissions mode of operation.

AUTO-CALL CAPABILITY

The auto-call capability can be ordered as an integral part of the modem, or it can be ordered as a separate device supplied by the telephone company. Either way, it is needed when it is desired to have a computer or intelligent terminal dial the locations on a switched network in accordance with a table of telephone numbers stored in memory. The digital signals of the computer are converted to dial tones for use on the regular voice dialing network. When a connection with the remote location has been made, the control of the line is transferred to the originating modem as it goes into the data mode.

Some modems are constructed to serve a variety of applications such as leased line two-wire or four-wire, dial-up, point-to-point, and multipoint. Others, in order to reduce the cost of features not needed for some applications, are set for restricted use such as transmit only and receive only.

SCRAMBLING

Modems can incorporate a technique, called scrambling, that helps to maintain the power of the signal at a more constant level. The transmitting modem rearranges the data into a more random pattern that removes undesirable sequences of repeated data that could lead to interference with other data channels on the same physical communications line. The receiving modem then decodes the signal to restore it to its original data pattern. Some standards like CCITT V.27 and V.29 recommend scrambling at higher speeds to combat the risk of transmitting long strings of 1s and 0s.

CABLE LENGTH

Though the EIA and CCITT standards call for a 50-ft limitation on the length of the cable connecting the terminal (or computer or front end processor) to the modem, some modem manufacturers have extended the limitation distance to 200 ft or so.

SPEED ON LEASED LINES

The previous maximum speed of 9600 bps on a leased line has been expanded to 14,400 bps and even more by some modem manufacturers. Line conditioning at a nominal additional cost would normally be required. Also, 19,200-bps modems have been offered for some time, but that line speed, being nonstandard, could require a specially engineered leased communications line that is economically impractical. The way around the special line problem is to utilize two standard 9600-bps leased lines.

DATA THROUGHPUT

Modem manufacturers are constantly working on new ways to improve their products, so it would be advisable to check the marketplace for current offerings prior to making a selection. Some manufacturers have employed the technique of data compression, which is claimed to double the throughput of a modem ranging up to 19,200 bps over a voice-grade telephone line. This could be built into the modem, or it could be a separate box that is positioned between the terminal and the modem. The data compression algorithm analyzes any ASCII- or EBCIDIC-coded incoming data and then converts the character to a shorter subcode based on its relative frequency in the data stream. For example, if the letter A is used frequently in the stream of data, the EBCIDIC eight-bit character (100000001) is converted to a new single-bit subcode. The modem or separate device will also operate as a multiplexer: it will handle four separate channels, provide full duplex operation, and perhaps even support X.25, SDLC, and IBM 2780 protocols.

Another approach used by some manufacturers to increase the modem throughput speed is the implementation of a newly available modulation technique, called Trellis coding, with forward error correction. In this approach the signal-to-noise ratio is reduced to make the line seem about 20 percent quieter. The claims are for speeds of 9600-bps full duplex transmission over dial-up phone lines and 16,800 bps over leased lines. Some of the modems are designed to slow down in speed if line conditions deteriorate and then go back to full speed when the poor condition disappears.

Interface Standards between Terminals and Modems

The Electronic Industries Association (EIA) is an organization in the United States that has set up interface standards for manufacturers and purchasers of electronic products. The International Telegraph and Telephone Consulting Committee (CCITT) is a similar organization that specifies communications standards internationally. In order for different manufacturers' equipment to be compatible and function correctly, it is necessary to abide by a predetermined set of standards. Unfortunately, there are multiple standards to do the same job, so we must be aware of what is available and what is being offered by the various vendors that we tend to mix as we select our terminals, modems, etc.

Some of the standards apply to balanced circuits and some to unbalanced circuits. Balanced circuits provide for the signal to travel along one conductor and return along another as with regular telephone line twisted pairs. Unbalanced circuits use a common return ground for several circuits, which is subject to more interference. In addition to the EIA and CCITT standards described below, there are U.S. Military Interface Standards that are a little more stringent. Most equipment that operates on the other two standards will also operate successfully on the military standards. The military standards are given in Appendix A.

EIA Standards

The EIA standards are primarily a set of electrical standards that control the relations of the operations between the modem, or data communications equipment (DCE), and the terminal, or data terminal equipment (DTE). The most popular standard in use today is the RS (Recommended Standard) 232C for serial binary data interchange, which is equivalent to CCITT standard V.24. It accommodates data speeds up to 20,000 bps over cable lengths up to 50 ft, which may be exceeded in actual practice. A binary 0 is represented by a positive voltage of 5 to 25 volts (V), and a binary 1 is represented by a negative voltage of -5 to -25 V. The connector itself is a 25-pin connector with from 3 to 25 of the available pins in use, depending on the particular application. Descriptions of the pins and their functions are given in Appendix B. The connector is shown in Fig. 4-7.

A newer EIA standard, RS-449, was implemented to provide all of the capabilities of the older RS-232C plus some additional circuits. It was hoped that the newer standard would replace the RS-232C, but the migration has been very slow. The RS-449 can be used as a 37- or a 9-pin connector. Also, it increases the maximum cable length from the terminal to the modem and has better diagnostic capabilities. In physical

Fig. 4-7   EIA RS-232C connector and cable (*Black Box Corp.*)

size, however, the connector is a little cumbersome. Other less popular EIA standards are listed below:

- RS-422A is designated to handle higher speeds over longer cables. It can accommodate 100,000 bps over a distance of 4000 ft.
- RS-423A handles data speeds of 3000 bps over cable lengths of 4000 ft, or 300,000 bps over 40-ft cables.
- RS-357 applies to analog facsimile equipment used for data transmission.
- RS-366 concerns the interface between a modem and automatic calling equipment.
- RS-408 is a standard for two interfaces between numerical control equipment (tape reader, etc.) and a serial-to-parallel converter.

CCITT Standards

A very popular standard set up by the CCITT is the V.35, which is used for the electrical control relations between the terminal and the modem for 56,000-bps lines. A 34-pin Winchester connector is used, and the pin functions are very similar to those previously described for the RS-232C. Other CCITT standards are briefly described in Appendix C.

Interface Operation Procedure

Two of the possible procedures that are followed during the operation of the interface circuits are described below. They are the more complicated dial-up line and the least-complicated point-to-point full duplex leased line.

The interface procedure for a switched line is performed as follows:

- The RECEIVE DATA circuit of the local originating modem is on MARK–HOLD and the DATA TERMINAL READY circuit is ON when the remote (distant) modem is called either manually or via an auto-call arrangement.

- The RING INDICATOR circuit goes ON at the remote modem, where the call is usually answered automatically, and the modem receives a DATA TERMINAL READY signal from its terminal. The RECEIVE DATA circuit goes to MARK–HOLD status. The remote modem then turns on DATA SET READY, lighting the DATA push button on the modem. After waiting 1.5 s, the modem issues an F2M tone to disengage all echo suppressors on the line and start the phone company's billing procedure for the call. If the call was manually originated, the individual at the local end will hear the data tone and press the DATA push button on the modem to go to DATA SET READY, which will light the push button. Then, 150 ms after receiving the F2M tone from the remote modem, the RECEIVE DATA circuit of the local modem goes to a NON-HOLD condition for 1.5 s and then sends out an F1M tone. After another 265 ms both the CARRIER DETECT and CLEAR TO SEND circuits are turned ON, making the transmit circuit available to the terminal.

- One hundred fifty milliseconds after the F1M tone is received at the remote modem, the MARK–HOLD is removed from the RECEIVE DATA circuit. CARRIER DETECT and CLEAR TO SEND are then turned ON after 265 ms, making the TRANSMIT DATA circuit available to the terminal. The remote modem then transmits a protocol-coded character to indicate that it is ready to receive data.

With a full duplex point-to-point line we have the simplest arrangement, which often requires only three of the interface circuits. The CLEAR TO SEND circuit can be turned ON when the modem is ready to send data, or there is the option to require REQUEST TO SEND prior to initiating the CLEAR TO SEND. Other than this, the only circuits needed are TRANSMIT DATA, RECEIVE DATA, and the SIGNAL GROUND.

Acoustical Couplers

For low-speed (up to 1200-bps) asynchronous transmissions an acoustical coupler can be employed. It consists of a small box with two foam cups for the telephone handset. The device converts the terminal's digital signals to audible tones which can be transmitted over any touch-tone telephone. There is the option to *hard-wire* the acoustical coupler device to the phone line and thus eliminate the possibility of background noise. Of course, the acoustical coupler is a very popular adjunct to small, portable terminals that can be used by salesmen at any motel. There is also a pad that can convert a dial phone to a touch-tone phone when that service is not available (see Fig. 4-8).

Modem Eliminators

A second way to get around using modems for distances of perhaps less than 100 ft, usually in the same room or building, is to install a *modem eliminator*. It could be for a terminal, printer, or minicomputer that cannot hook directly into the mainframe computer's channels but must operate as a remote communications device. This box looks like a

Fig. 4-8 Acoustical coupler (*Multi-Tech Systems, Inc.*)

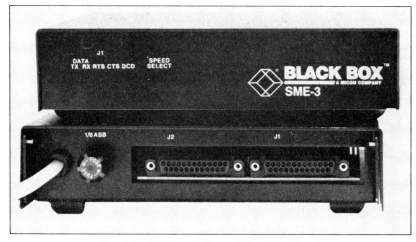

Fig. 4-9 Modem eliminator (*Black Box Corp.*)

modem, but it is smaller and has limited functions that do not include modulation. Only one modem eliminator is needed on a single line (see Fig. 4-9).

Line Drivers and Limited-Distance Modems

Both line drivers and limited-distance modems are other means of avoiding the expense of regular modems where short distances are involved; the latter are a little more sophisticated and permit a little longer run. A line driver can be used at speeds up to 19,200 bps and over distances up to a few miles. A synchronous line driver contains the transmit and receive clocking circuits and the control signals. Originally the limited-distance modem was restricted to something like 5 mi of hard copper wire without telephone company coils and repeaters. Some limited-distance modems that have become available more recently operate well for up to 50 mi on unconditioned leased phone lines (No. 3002) at speeds up to 4800 bps. Unlike the modem eliminator, both line drivers and limited-distance modems require a device at both the transmitting and receiving ends of a communications line (see Fig. 4-10).

Fig. 4-10 Limited-distance modem (*Black Box Corp.*)

Splitters

A splitter is a box about the same size as a modem that is used for multidropping terminals in the same building location. If multiple terminals are to share the same leased line, we normally assign each of the terminals its own polling address and create a station drop at each terminal location by inserting a modem in the line at that point. This approach is efficient if the terminal locations are miles apart; but if they

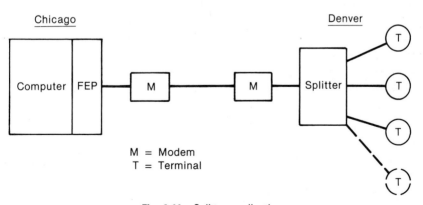

Fig. 4-11 Splitter application

are close together, as in the same building, it would be less expensive to install a splitter device which could provide for, say, four terminal drops.

The splitter assigns a polling address to each of its local ports, and any device attached to one of the ports, say, a CRT display or printer, will receive all of the data associated with that particular address. Figure 4-11 shows the connection of a splitter to multidrop three terminals at the same location.

5

Multiplexers and Concentrators

The *concentrator* is often confused with the *multiplexer*. Both devices can be used to connect multiple lower-speed terminals to a single high-speed line (or a few such lines), but there is a difference. A concentrator actually concentrates several communications lines into one faster line (or a few faster lines), usually to a data center. A multiplexer divides one communications link so it can be shared by several different terminals. Figure 5-1 shows an example of each for comparison. Note that two multiplexers are always required on a communications link, whereas only one concentrator is normally used. With a concentrator, the total input and output rates can differ, since the buffer memory can hold the excess data.

Concentrators

In Fig. 5-1, the 10 data stations shown in the concentrator network could be scattered within a 25-mi radius around the San Francisco area, the concentrator could be located in the center of San Francisco, and the computer center could be located in Los Angeles. The economic advantage here is to pay for only one of the expensive long-haul lines from the Los Angeles computer to the center of San Francisco, where each separate data station location has its own inexpensive short line to the centrally located concentrator. The long-haul line would normally be a faster line, such as 9600 bps if analog or 56,000 bps if digital. The multiple shorter connectors would normally be lower-speed lines that probably would not be used all at the same time.

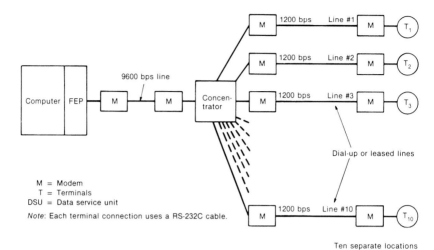

(a) Concentrator network

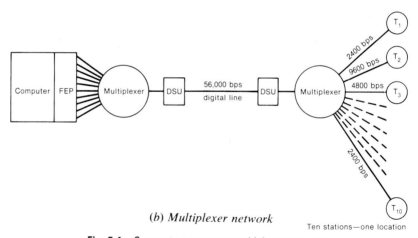

(b) Multiplexer network

Ten stations—one location

Fig. 5-1 Concentrator versus multiplexer networks

A concentrator must have a memory buffer to temporarily hold any overflow in case too many data stations transmit at the same time and overload the single 9600-bps line to Los Angeles. A concentrator could perform some other memory logic functions such as data editing, data storage for subsequent transmission, and protocol conversion. The protocol conversion function could be advantageous in providing a

Fig. 5-2 Concentrator (*Infotron Systems Corp.*)

more efficient protocol for the long-haul line back to the data center. Figure 5-2 shows a typical concentrator.

Multiplexers

A multiplexer is not normally concerned with lower-speed communications lines; instead, it is usually used with a higher-speed communications line (or a few such lines or a link involving microwave, satellites, etc.) that is to be shared by multiple lower-speed data stations. The data stations are usually located in the same building within 50 ft of the multiplexer unit and connected to it with EIA RS-232C type cables, as shown in Fig. 5-1.

The principal use and economic advantage of the multiplexer is to be able to utilize the full bandwidth of a higher-speed communications link continuously. The total dollar savings, of course, would have to exceed the cost of the multiplexer equipment needed at each end of the communications link. Figure 5-3 shows a typical multiplexer.

If desired, a multiplexer can accommodate lower-speed communications lines as an extension of one of the multiplexed channels to provide a tail circuit on any of the channels for a data station over 50 ft from the multiplexer box itself. The arrangement is basically the same as that depicted in Fig. 4-5.

Multiplexers vary widely in capacity and sophistication. The simplest and earliest example is a modem that splits a communications line into two channels, the second being the *side channel*. Frequency-division multiplexing was used to accomplish the splitting. It was later expanded to permit the modem to split a single analog telephone line into up to four channels as described previously. Then came the multiplexers (time-division multiplexers) that were constructed as separate devices and attached behind the modem. They normally would split a single higher-speed line into something like 12 separate channels at various optional speeds. Then the number of intermittently used terminals that could be accommodated on one line was increased considera-

Fig. 5-3 Multiplexer (*Infotron Systems Corp.*)

bly by a new method called statistical multiplexing. These three methods of accomplishing multiplexing will be described in more detail below.

The larger and more sophisticated multiplexers not only provide for a greater capacity of input lines and output channels but also expand the functions—often to include diagnostic aids and statistical line traffic information. These multiplexers are usually the statistical variety that handle both asynchronous and synchronous communications. They often include some additional logic to increase data block size, operate in full duplex, employ a more efficient line protocol, and even "fake out" polling to create an efficient transmission out of what would otherwise be a poor application for satellite communications.

The three methods of multiplexing are shown in Fig 5-4, which depicts the transmissions as being analogous to railroad box cars used to transport data.

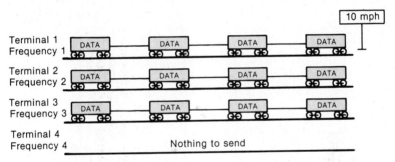

(a) *Frequency-division multiplexing—low speed, multiple paths*

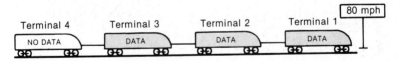

(b) *Time-division multiplexing—high speed, one path*

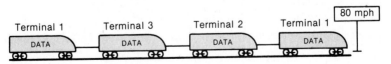

(c) *Statistical multiplexing—high speed, one path, no empty cars*

Fig. 5-4 Multiplexing comparisons

Frequency-Division Multiplexing

The earliest form of multiplexing, by frequency division, is based on the fact that a single conductor wire can provide paths for signals of several different frequencies. For instance, a single phone line can provide paths for 12 separate frequency transmissions of 3000 cps each. Frequency-division multiplexing is used quite extensively in the transmission of voice (analog) signals as well as data (digital) signals. Each frequency path occupies its own cylinder position in the wire.

An early example of frequency-division multiplexing for the transmission of data is the use of the *side channel* of a regular leased phone line for a teletype terminal at up to 300 bps while the main channel is used for an RJE printer at 2400 bps. Most leased line modems provide a side channel feature at very little additional cost. The side channel can also be divided into two 250-bps channels, one for low-speed data transmission and perhaps the other for a network control test channel. A similar application is the splitting of communications lines into perhaps four 2400-bps channels by a special multiplexing feature provided with some modems.

A current and very popular application of frequency-division multiplexing is the operation of a two-wire telephone line to look like a four-wire full duplex facility. The 212A-type modems accomplish this by designating one frequency as the first two wires and another frequency as the second two wires.

Frequency-division multiplexing is the least complicated and least expensive of the three multiplexing methods. As would be expected, it has the disadvantage of being the least efficient in regard to total throughput.

Time-Division Multiplexing

Time-division multiplexing came into use next, and it provided the advantage of running the communications line at its full speed. There is only one full-bandwidth path, which must be shared by all of the terminals using the communications link. If there are four terminals, then every fourth frame (or box car in Fig. 5-4) is reserved for the fourth terminal whether or not it has data to send. This form of multiplexing requires programmed logic, but it has proved to be more cost-effective than frequency-division multiplexing in the larger applications.

Time-division multiplexing was used in the example in Fig. 5-1. There the multiplexer divides the 56,000-bps digital four-wire leased line into five 9600-bps individual channels, or most any combination of lesser bit rate channels that totals a little less than 56,000 bps. The installation of this type of multiplexer is relatively simple. The four wires of the digital communications line are connected to the modem (or data service unit for digital lines), to which the EIA RS-232C or equivalent cable from each terminal is attached. Each of the eight or more channels available is assigned a bit rate by means of a switch or small patch wire.

There are two different general approaches to designing a time-division multiplexer. One is to buffer a bit at a time, and the other is to buffer a full character at a time. The real advantage to character buffering is with asynchronous transmissions in which the one start and the two stop bits can be deleted from the actual transmission over the communications line. This will increase the speed of the data being sent because it is 27 percent more efficient. The character buffer approach increases the cost of the multiplexer, and it can not be cost-justified for synchronous transmissions. Character buffering creates a problem for some systems, e.g., Telex, that require the smaller transmission delay inherent in bit buffering.

Statistical Multiplexing

Statistical multiplexing is the most recent innovation. It is similar to time-division multiplexing in that data is transmitted at the full line speed of 9600 bps or so for an analog line and 56,000 bps for a digital line. It differs in that a frame position is not reserved for each participating terminal; so if a terminal has no data to send, its frame can be used by another terminal that has data. As noted in Fig. 5-4, no "empty box cars" are sent.

Statistical multiplexing was originally designed for low-speed asynchronous transmissions only, but it has more recently been enhanced to include synchronous transmissions also. Since this approach involves the sharing of a single higher-speed link, it becomes most effective for terminals that send data only intermittently. Thus we see that a printer that runs continuously might do as well under frequency-division multiplexing in which a path is reserved, and it would not block out the other intermittent terminals. The blocking problem could be resolved with a priority system under statistical multiplexing, but the real throughput advantages diminish when terminals transmit continuously.

6

Front End Processors

A front end processor is a computer that acts as an interface between the mainframe computer (host) and its associated communications lines to remote terminals or other computers. It was originally called a transmission control unit and more recently given the name communications controller. Figure 6-1 shows a typical front end processor. The functions of a front end processor vary with the degree of sophistication needed and, of course, the cost.

The earlier devices, such as the IBM 2701/2703, were called transmission control units and were limited to functions 1 through 4 listed below. Later came the communications controllers, such as the IBM 3705, that were ultimately called front end processors. The newer devices add most of the remaining functions listed, and a few qualify as the ultimate in off-loading cycles from the mainframe computer in that they store data on disk.

Figure 6-2 shows an example of a multinode three-city data communications network in which any data station in the network can access any application program in either of the two computers. Note that the San Diego location does not have a computer but does have a group of user data stations, which can be located in the same building as the front end processor or miles away by using communications lines and modems.

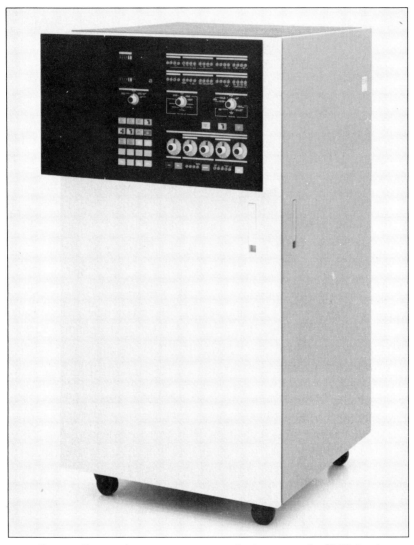

Fig. 6-1 Front end processor or communications controller (*IBM Corp.*)

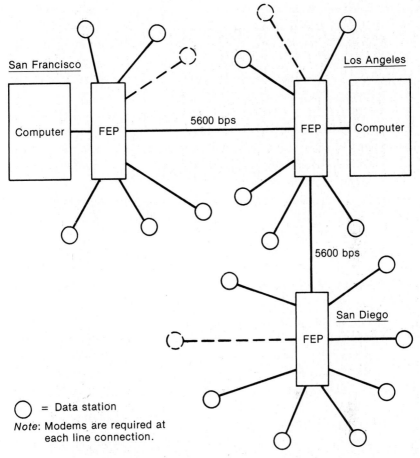

San Francisco

Los Angeles

5600 bps

5600 bps

San Diego

◯ = Data station

Note: Modems are required at each line connection.

Fig. 6-2 Multinode communications network

Functions of a Front End Processor

Some of the more common functions of front end processors are the following:

- Carry out the communications line discipline, which includes adding control characters for outgoing traffic and stripping them off for incoming traffic.
- Contain hardware devices called communications line adapters to send and receive signals on each communications line.

- Take incoming bits that arrive serially and assemble them in a memory buffer into characters of usually from five to eight bits and also disassemble characters into serial bits for outgoing data streams.
- Handle error detection and correction in accordance with the protocol requirements.
- Perform any code conversion needed, such as ASCII to EBCDIC (needed by the computer).
- Perform diagnostic line tests and accumulate traffic statistics.
- Automatically send signals to dial out when so requested by the communications software.
- Take over the complete polling operation for multipoint terminals, including the scanning of the polling list in the front end processor's own memory.
- Accumulate incoming and outgoing data in a buffer to make up a word (32 bits, etc.) or even a block (200 characters, etc.), in order to cut down on interruptions to the mainframe computer.
- The newer software-controlled front end processors can provide application switching whereby the same terminal (e.g., CRT display) can access several different communications monitors, each with its own applications on the associated mainframe computer (e.g., IMS, TSO, CICS, etc.). There are two ways to do this. One is to go all the way with the full software package such as IBM's VTAM in the host computer. Other front end processor vendors offer the application switching without all the expensive mainframe software, but the trade-off is in saving fewer mainframe computer cycles.
- Support a network of multiple front end processors around the country whereby any terminal in one city can communicate with any terminal in another city.
- The more recent front end processors usually have printer-keyboard consoles that provide capabilities such as switching line adapters (for troubleshooting) and clearing and resetting lines.
- Instead of requiring an individual connection between the front end processor and the mainframe computer for each communication line (separate subchannel addresses), as few connections as one can be utilized to conserve the limited subchannel addresses available to a mainframe computer.

• Some front end processors store data on disk (or tape) to limit the interruptions to the mainframe computer even further and perhaps perform any data transfers at off-hours. Editing, date stamping, and other activities also can be performed to save cycles on the mainframe computer.

IBM 3705 (and 3725) FEP under NCP and SNA

A discussion of front end processors would probably be incomplete if we did not include a description of IBM's 3705 as it is used in the Network Control Program (NCP) with Advanced Communication Function/Virtual Telecommunications Access Method (ACF/VTAM) with the overall Systems Network Architecture (SNA). The Multi-System Networking Facility (MSNF) is also required in multiple host networks.

When in our discussion we refer to the IBM 3705, the same information will be applicable to the newer IBM 3725 front end processor, as well as the IBM 3704, which is a smaller-capacity unit. Many more of the IBM 3705 units are installed around the country today, but they probably eventually will be replaced by the more cost-effective IBM 3725.

It should be noted that the IBM 3705 itself is a general-purpose front end processor and can accommodate a wide variety of different kinds of communications lines and data stations. However, in order to take advantage of all the features of a full-blown SNA network, we must remember that the original design was conceived with only certain IBM data stations in mind. Other vendors have manufactured equipment that is termed compatible with the SNA requirements, and IBM itself has since added modifications to handle teletype terminals and even an interface to X.25 protocol. This latter feature provides a gateway to an X.25 network but does not permit SNA to be incorporated into an X.25 network acting as an actual node. IBM has just recently introduced the 3710 Network Controller, which attaches to the 3705 or 3725 and acts as a cluster controller protocol converter. Up to 31 terminals can be connected to an SNA network under asynchronous, binary synchronous, or X.25 protocols.

IBM 3705 AND NCP

The IBM 3705 Communications Controller is a programmed transmission control unit that can be attached directly to a host computer via a channel adapter or can be located many miles away as a remote communications node. The increased capability of taking over more line control functions is provided by the Network Control Program (NCP) that is executed in the 3705 itself. It relieves the host computer of much of the teleprocessing responsibility. The 3705 can be run in the emulation mode like an IBM 2703 transmission control unit, or under NCP if the SDLC protocol and SNA are desired, or with both NCP and emulation modes by means of a Partitioned Emulation Program (PEP). With the latter, some lines can be run under NCP and some in the emulation mode, or the same lines can be alternated between the two modes via a command from the access method program in the host computer. When operation is under NCP, only a single subchannel address is needed to communicate with the host computer as compared to a separate address for each communications line under the emulation software package.

The NCP program specifies the numbers, types, and configurations of terminals to be attached to the IBM 3705. It describes the IBM 3705 itself, specifies any options, identifies the interaction of the host computer operating system with the NCP (buffers, block sizes, etc.), and describes lines, line groups, paths, and terminals. The NCP takes over most of the control of communications lines from the teleprocessing access method; that is, it saves machine cycles on the host computer. The main functions of the NCP are listed below.

- Polls and addresses teleprocessing units on multipoint communications lines.
- Dials and answers stations over a switched communications network.
- Recognizes an interrupt when either a character or a bit (depending on the type of scanner selected) arrives over a communications line. The program moves the character or bit into a buffer.

- Inserts control characters at the beginning and end of each block of data when transmitting to a station and deletes them when receiving from a station.

- Determines to which remote controller a block of data is to be sent in a multinode network.
- Controls message traffic between a local controller (central computer) and remote controllers (decentralized computers). This includes transmitting program load modules (NCP software programs) to remote controllers and passing memory dump data from remote controllers to the host access method program.
- Translates any binary synchronous or start-stop codes into EBCDIC (host computer code).
- Allocates buffers for data storage on a dynamic basis.
- Reduces line speed to half the usual rate if throughput degradation is detected.
- Keeps a record of the front end processor hardware and program problems as well as permanent line errors and operational statistics. Figure 6-3 describes the data communications relations of the host computer, front end processor, and remote terminal.

In all fairness, it should be mentioned that other vendors do offer front end processors that are comparable to the 3705 and can operate under SNA with SDLC and VTAM. There might be a little concern whether the compatibility is 100 percent, which the vendors do claim is

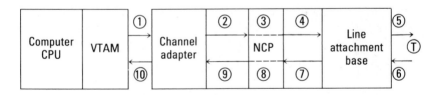

① CPU sends data to FEP.
② Channel adapter notifies NCP as data arrives.
③ NCP processes data for terminal (T).
④ NCP activates scanner for transmission of data.
⑤ Data is transmitted over communication line to terminal.
⑥ Terminal transmits data to FEP.
⑦ Scanner notifies NCP as data arrives.
⑧ NCP processes data for CPU.
⑨ NCP activates channel adapter for transmission.
⑩ Channel transfers data to CPU.

Fig. 6-3 Operation of IBM 3705 FEP

true, and any new modification releases may be delayed a while. On the other hand, some other vendors appear to offer greater flexibility and lower prices.

Systems Network Architecture

SNA is an overall planned total system structure for a data communications network. It has earned the status of the world's leading architecture for private networks with over 5000 installations and continues with a steady growth. Some of the advantages are improved response time, decreased line costs, decreased host computer load, and improved availability of critical functions in the event of a component failure. Some of the additional features are a distributed function, device independence, attachment independence (i.e., data stations can be added any time), and configuration flexibility.

SNA deals with three types of network-addressable units (NAUs). The first is the *system service control point* (SSCP), which is a program residing in the host computer that, among other things, provides network services such as handling operator requests and establishing sessions. The second is the *physical unit* (PU), which is the actual physical node and data station. The third is the *logical unit* (LU), which is the port through which the end user accesses the system. A *session*, which is a formally bound pairing between two points (NAUs), can be LU to LU, LU to SSCP, or PU to SSCP.

SNA (system architecture) is divided into a set of three logical layers that permit modifications without affecting other layers and provide for interactions between functionally paired layers in different units. Thus, individually separated layers that are alike can communicate with each other without going through the other two layers. The three layers are described below.

> Transmission subsystem—controls the routing and movement of data units between origin and destination and permits the sharing of the various paths throughout the network.
> Functional management layer—controls the presentation of information from one application layer to another. A data flow protocol assists the user in controlling the flow of requests and re-

sponses. Data is transformed as needed by the use of presentation services and logical unit services. These services include support for the user's application programs and hardware terminals.

Application layer—controls the user's application program functions and keeps it distinctly separate from the data transmission functions.

The transmission subsystem is composed of three major elements:

Transmission control element—permits a user to access the transmission subsystem with requests, responses, etc., by invoking the *connection point manager, session control,* and *network control* components.

Path control element—manages the routing of basic information units (BIUs) through the network.

Data link control element—manages an individual data link.

The transmission subsystem moves request/response units (RUs) from one network addressable unit (NAU) to another. One NAU will pass an RU to the transmission control element for a session with another NAU, where a request/response header is added to form a basic information unit (BIU). The BIU is then passed to path control, where a transmission header (TH) is added to form a path information unit (PIU). The PIU is then passed to data link control, where link control information is added to form a basic link unit (BLU) and then send it over the data link to the next node. Figure 6-4 shows the relations of the BIU, PIU, and BLU as they make up a frame. In one

3 bytes	2–10 bytes	3 bytes	variable	variable	3 bytes
Link control information	Transmissions header (TH)	Request/ response header (RH)	Functional management header (FMH)	Request/ response unit (RU)	Link control information

Basic information unit (BIU) →
Path information unit (PIU) →
Basic link unit (BLU)

Fig. 6-4 SNA transmission subsystem frame

new and interesting concept employed here, called *pacing,* each connection point manager function regulates the rate at which it receives data flow requests so that the sending location is not permitted to transmit data any faster than it can be received.

SNA was originally designed and released back in late 1974, based on a master/slave relationship between the host computer (the master) and associated terminals (the slaves). The host computer controlled all of the network system activity down to the functions of the application program residing in the host computer. This lack of a peer-to-peer relationship between the host computer and the various terminals was finally relieved in mid-1982 by the introduction of a software package called Advanced Program-to-Program Communications (APPC), or Logical Unit (LU) Version 6.2. This new feature provided the same protocol for communications between various application programs from host to host, terminal to terminal, or terminal to host.

The original configuration and then all subsequent changes to an SNA network must be implemented by means of software program specifications. This programming operation turns out to be a complex procedure requiring considerable time and effort. This ''system generation process'' has been shortened by a factor of four to eight times through the introduction of a software package called Advanced Communications Function/System Support Programs (ACF/SSP), Version 3.0.

SNA/X.25 Communications Network

The design of a fairly sizable data communications network containing multiple front end processors requires some important considerations and objectives that the network designer would want to accomplish. Of course, the costs of implementation and operation are important, but even more important are such factors as reliability, speed, response time, flexibility, and overall convenience of service to the ultimate users.

The X.25 packet switching network provides many advantages to the network designer. Polling overhead is reduced from the primary network. With virtual terminal mapping a terminal can select its receiving application prior to host connection and thus eliminate the need for

SNA's System Service Control Point (SSCP) switching, resulting in what is called dial-in access. The addition of individual user terminals or even host computers is also accomplished more easily. Circuit costs can be reduced by multiplexed session traffic and because there is no need for long-haul polling. Packet networks also provide for alternate communications paths in case of a circuit breakdown.

SNA also has its advantages. It can handle bulk data traffic, such as remote job entry (RJE) print stations more efficiently. It can handle front end processor–to–front end processor (FEP–to–FEP) communications, which could be a problem with packet switching networks. On the other hand, SNA does have a station address limitation which could affect very large networks of over 30,000 terminals.

EQUIPMENT MIX

Because of the evolution of data processing facilities of companies in the United States, plus the desire to mix equipment offered by a variety of vendors, the average company finds itself with a mixture of unlike terminals and computers and a desire to take advantage of the capabilities of more than one type of data transmission system protocol. Most companies start out with asynchronous (teletypes, etc.), 2780/3780 bisynch (RJE printers, etc.), and 3270 synchronous (clustered CRT displays, etc.) terminals and would like to retain that paid-for equipment. At the same time, a progressive company wants to take advantage of the newer bit-synchronous protocols when possible and particularly when any new terminals are to be installed. Two of the main contenders in this area are IBM's SDLC protocol and the internationally popular X.25 protocol. The former is well entrenched in this country and has a very impressive list of using companies. The latter has gained its contending position primarily through its use by the public packet switching networks and its availability for a private network not limited by the restrictions of certain equipment as required under IBM's SNA architecture. The following are some of the desirable features of a major data network:

- Connects multiple front end processors around the country into a multinode network.
- Permits any terminal to access any application program on any of the multiple host computers.

- Permits any terminal to communicate with any other terminal on the network.
- Allows asynchronous (start-stop), synchronous (3270), RJE (bi-synchronous remote job entry), and X.25 terminals to share the same communications lines (but separate channels) to reduce line costs.
- Implements SNA for the private network consisting of 3270 clustered displays, RJE printers, etc.
- Adds an X.25 protocol capability to afford communications with a public packet switching network and also efficiently handle X.25 minicomputers and terminals.
- Attempts to maintain the efficiency of direct SNA/SDLC transmissions for high-volume FEP–to–FEP communications rather than confine all communications to X.25 with the excessive overhead of wrapping data in an inner SDLC envelope and an outer X.25 envelope. There should also be the option to switch to the X.25 network route if the SDLC lines get into trouble.
- Keeps the polling of terminals down to only the local front end processor with which the terminals are more closely associated rather than require every front end processor on the network to poll every terminal. Thus only useful data is sent out on the network trunks when possible.

A diagram of a data communications network that encompasses all of the desirable features described above is shown in Fig. 6-5.

SEPARATE FRONT END PROCESSORS

If a very large network is being designed, the above-described principles can be further expanded to the extent that separate front end processors are employed for X.25 packet switching networks and the SNA network. This approach would normally be used for extremely large networks in the area of 25 host computers and perhaps 30,000 individual terminals. Here again it was found that neither technology could fully satisfy the requirements of the network, so the combination was implemented and called Extended Network Architecture (XNA). The basic idea of XNA is to provide each host computer with a separate front end processor for its packet switching network and then connect all host computers at a particular node or data center to another

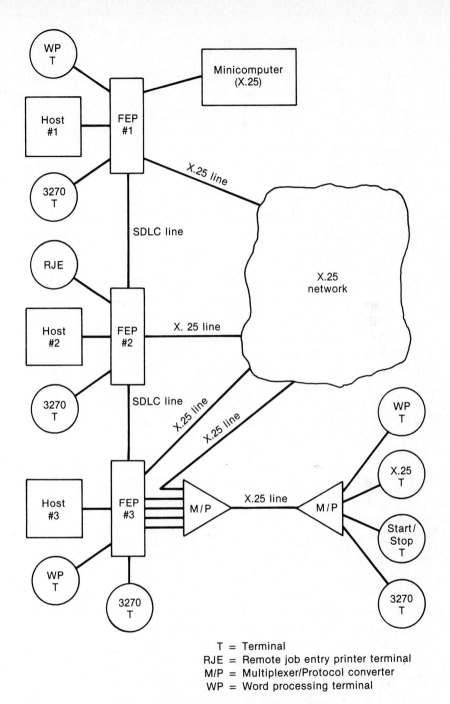

T = Terminal
RJE = Remote job entry printer terminal
M/P = Multiplexer/Protocol converter
WP = Word processing terminal

Fig. 6-5   SNA/X.25 communications network

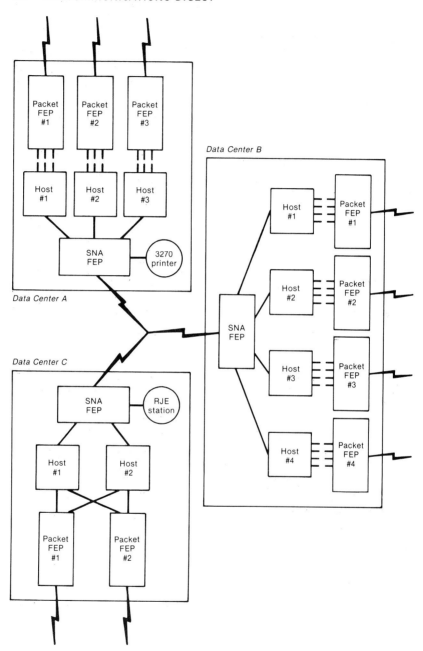

Fig. 6-6 XNA communications network

"switch" front end processor that provides access to the common SNA network. The packet switching network contains all of the terminals, and the SNA network contains all of the host computers and RJE print stations. The only limitation of this XNA arrangement is that only eight host computers can be attached to one front end processor and its Network Control Program (NCP). If there are over eight at a single data center location, multiple XNA nodes are required at that location (see Fig. 6-6).

7

Terminals for Data Communications

Literally thousands of different terminals are available on the market today for the transmission and receipt of data over communications links. Needless to say, it is important to select the proper terminal to suit the needs of a business application. Here we will attempt to identify the various types of terminals by category and discuss the merits of each.

CRT Displays

The CRT display shown in Fig. 7-1 is by far the most popular type of data entry terminal available today. It is used primarily for interactive communications in which a request is sent via the display keyboard and an answer is received on the screen within, say, five seconds or so. Of course, data can be entered continuously throughout the day by a data entry clerk with no more response than that the message is accepted. Normally, however, the screens on the display are formatted to assist the operators, and that is one of the big advantages of the display terminal.

The display terminal has a keyboard, which can be arranged like a regular typewriter keyboard or as a data entry or 10-key pad configuration. The number of characters available on the screen can vary, but the most common is 24 lines of 80 characters each for a total of 1920. Color also is available at an increase in equipment cost, plus the additional processing time on the associated control unit.

A CRT display requires data communications logic to interface with a communications line. A single stand-alone display usually has

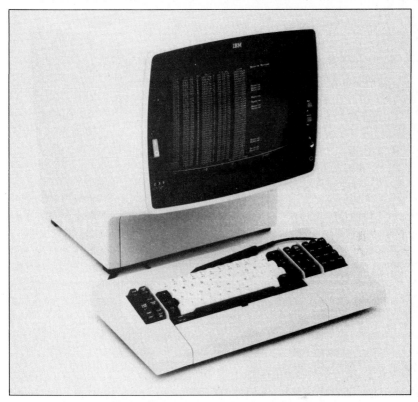

Fig. 7-1 CRT display with keyboard (*IBM Corp.*)

this logic built into the display cabinet as an extra logic board. Most locations require large numbers of displays, plus associated character printers. Therefore, they are set up to connect to a separate cluster controller, which normally provides for up to 32 ports for displays and printers to be attached (see Fig. 7-2). The earlier terminals can usually be up to 2000 ft from the control unit and the more recent ones 5000 ft. Each is connected by a ¼-in coaxial cable with special connectors.

INTELLIGENT VERSUS UNINTELLIGENT TERMINALS

CRT displays have been categorized as "intelligent" versus "unintelligent," or "dumb," the reason for the division probably being the wide variation in specific functions available in the displays being offered today. We can point out some of the functions that would tend to

Fig. 7-2 Cluster controller (*IBM Corp.*)

put a display in the "intelligent" category. The ability to operate synchronously at full duplex would be a plus, as would be the capability of operating in the 3270 mode to gain all of the associated screen formatting advantages, etc. Of course, there are protocol converters, multiplexers, etc., that can add these features if placed in front of the display terminal. Another good feature would be the capability of being able to print any screen of data on an associated small character printer without first having to send the data back to the host computer. This is called the *copy function*. The criteria probably used most often to identify an intelligent terminal would be the capability of retaining its own screen formats and an editing function. The latter cuts down the data communications traffic load, improves response time, and saves processing cycles on the host computer.

GRAPHICS

Another feature of the CRT display terminal that is becoming increasingly popular is the capability of displaying graphics. This, of course, increases the cost of the display, but it is well worthwhile for many applications and particularly for computer-aided design/computer-aided manufacturing terminals (CAD/CAM). CAD/CAM systems are becoming a necessity for even small manufacturers, who can purchase the terminals, lease the telephone line, and pay for timesharing on a local computer center's CAD/CAM system application.

Teletype Terminals

Figure 7-3 shows a typical teletype terminal. It is actually a general-purpose terminal that operates like keyboard-send-receive (KSR) teletype devices. Terminals of this type originated for use on the TWX and TELEX message networks, but they became very popular as timesharing terminals whereby remote locations could access a distant computer to run business applications and create the associated computer programs.

The KSR-type terminals probably represent the least expensive way to send messages, data, and requests to a computer and get back

Fig. 7-3   Teletype terminal (*Teletype Corp.*)

printed documents. These terminals normally operate asynchronously at 300 to 1200 bps, and many of them permit full duplex capability by using the 212A-type modem. Diskettes, magnetic tape cassettes, memory buffers, and even punched paper tape devices can be utilized to enter larger amounts of data off-line and transmit at the full line speed rather than a character at a time with pauses in between.

RJE Line Printers

When long reports or large volumes of invoices, orders, etc. must be printed at a remote location, the remote job entry (RJE) printer is often utilized. It is very popular for use by programmers to enter jobs for compiling and testing programs, especially if the programming department is miles away from the computer. The RJE terminal usually consists of a fairly high speed printer of about 600 lines per minute with a card reader for entering jobs. A CRT display is sometimes added for the convenience of avoiding card punching to initiate job instructions (see Fig. 7-4).

One of the primary goals in the installation of an RJE station is to obtain the maximum printing speed out of a single leased or dial-up line. The terminals usually run at a maximum speed of 9600 bps on a

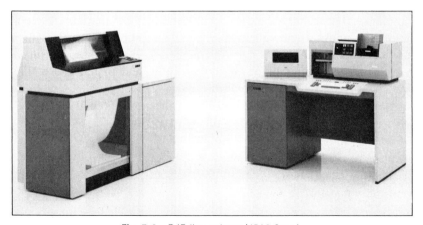

Fig. 7-4 RJE line printer (*IBM Corp.*)

four-wire leased line and 4800 bps on a dial-up line. Without any other help, the printers ordinarily will operate at 550 lines per minute at the 9600-bps line speed by using two buffers in the RJE station. The printing speed can be increased by adding a feature called *character compression*. The transmitting software can identify any duplicate characters over three in number; instead of sending all of the duplicates, it sends a three-character instruction code to, say, print 100 dashes. This becomes extremely effective when we think of all the blank spaces in an average document. Most RJE transmission software packages automatically omit trailing blanks in any print line being sent.

Another way to run the printer at a higher speed is to add disk or magnetic tape storage to the RJE station, take transmission directly onto disk or tape, and later run the printer at full channel speed. This has the disadvantage of processing the same data twice. Of course, a wider band line like 19,200 hps is another option for faster print speeds.

Distributive Data Processing Station

The devices shown in Fig. 7-5 are actually small computers situated at remote locations that are needed to run business applications locally but also are needed to communicate with the mainframe computer periodically. For instance, a small computer at a warehouse could receive orders from the host computer at headquarters while on-line with communications. Then the warehouse computer could print all of the shipping documents during the workday and update the inventory disk files as it processes shipments. At night the small remote computer could transmit its updated inventory file to the headquarters host computer for updating the corporate master inventory files, etc.

A typical distributive data processing station consists of a small computer processor, disk storage devices, a card reader, multiple CRT display terminals for data entry, a 40 to 400 line per minute or even faster printer, and perhaps a magnetic tape drive unit. The data communications control unit is usually built into the computer processor and then does not require a separate front end processor. If there is a considerable volume of transmissions between the two computers, a 56,000-bps digital communications line might be in order.

Fig. 7-5   Distributive data processing system (*IBM Corp.*)

Facsimile Transmission Units

In facsimile transmission the image on a piece of paper is scanned at the point of transmission and reconstructed at the point of receipt. The scanned photocell signal is modulated and converted to amplitude- or frequency-modulated signals, which are then demodulated at the receiving end. Printing at the receiving end is usually by an electrothermal process whereby the voltage applied to a stylus causes a flow of current that generates enough heat to "melt" the top white layer of the paper and expose the second darker layer and so create an image of printed or photographed information (see Fig. 7-6).

Facsimile devices can be used to send printer documents, drawings, pictures, etc. from one location to another. There is also the option to transmit data from a computer to a facsimile machine, when the two are compatible, or by using a facsimile public network. The real advantage here is that the tedious step of keying in data to be

Fig. 7-6 Facsimile transmission unit (*3M Corp.*)

transmitted has been eliminated. Almost any sheet of paper with information on it can be transmitted by simply inserting the paper in the machine, placing a regular telephone call to the destination location, and pressing a start button. Higher transmission speeds can be obtained by using leased lines, but because most facsimile machines are used for low volumes to a variety of locations, dial-up lines are usually employed. The speed with dial-up lines has progressed from 6 min per page to 3 min, and now, with digital facsimile, the time for a page is down to about 15 s. Full duplex facsimile also is available.

The operation is usually better if all the participants have the same vendor's equipment. There is some intervendor incompatibility, and a facsimile public network can be employed to get around the problem.

Character Printers

Character printers, in contrast to line printers, print a character at a time and thus have lower speed ratings of something like 40 to 180 characters per second. They are used for applications in which the

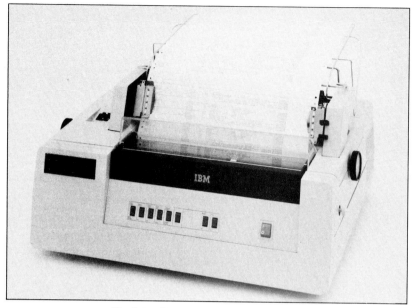

Fig. 7-7   Character printer (*IBM Corp.*)

volume is relatively low or the finished documents are not required immediately and can wait for the printing process to be completed. Character printers are usually quite noisy and so should be located in separate rooms or away from employees (see Fig. 7-7).

One of the popular uses for character printers is in conjunction with CRT displays. The printer can be attached to the same control unit and either print directly from a display screen (copy function), or it can print from the communications line as received from a host computer.

Word Processing Systems

The word processing system shown in Fig. 7-8 is a very popular office automation product that usually consists of a small computer to which are attached disk storage units, printers, and multiple displays. This is the computerized approach to automating the secretarial job. Instead of typewriters, CRT displays are used to permit easy correction and

Fig. 7-8   Word processing system (*IBM Corp.*)

rearranging of any keyed information. Any information typed previously is retained in a disk file for reuse or modification as desired. There are a lot of typing aids ranging from line justification to spelling assistance, depending on the degree of sophistication desired.

Most of the data manipulated by the word processor is for local use in an office environment, but facilities are usually available to permit the system to communicate with the local mainframe computer or a remote computer or another word processor. A word processor can be used as a data entry device and also as an electronic mail system computer. Character printers are normally used with word processing systems: one for formal letter paper and another for pin-feed in-house documents.

Portable Terminals

Portable data communications terminals have become very popular to fill a need, such as permitting a salesperson to check inventory and

Fig. 7-9 Portable terminal (*3M Corp.*)

initiate an order while he or she is on the road, or for a student or programmer to work at home (see Fig. 7-9).

The most common portable terminal could well be the small teletype or typewriter terminal consisting of a keyboard with a nonimpact type of page printer. Connection to a dial-up communications line could be made with an acoustical coupler or a direct wire phone jack. The acoustical coupler tone transmission method is normally restricted to a speed of 300 bps, but the speed can be increased to 1200 bps if an integral or separate modem is used.

Another popular portable terminal is a small CRT display. Like most portable devices, it is an asynchronous unit that employs the regular dial-up phone lines. The least expensive "portable terminal" of all is the touch-tone telephone. It can be used to transmit data by pressing the tone pad keys. Of course, the entry of data is limited to low-volume activity, but telephones could be useful for the right application. A tone or audio response from the data center could be used to acknowledge or reject any entry. Sending a tone requires the sending of a character back to the user. A special audio response unit installed at the data center could send back audio sentences concerning corrections required or information such as costs or calculation results, or it could even prompt the user with instructions on the next step of the procedure.

The vocabulary of an audio response unit is limited, but it is sufficient for most applications. Also, it can be used in conjunction with any terminal by switching back to the telephone instrument.

Personal Computers

Small, personal computers are rapidly finding their way into today's business world (see Fig. 7-10). Even if a company has large computers to handle all of the major business applications, there still appears to be a need for individual departments to be able to set up their own programs to handle some of the business applications peculiar to depart-

Fig. 7-10   Personal computer (*IBM Corp.*)

ment needs. Availability and control become important factors, especially in smaller office locations or private businesses that do not have access to a large computer.

Needless to say, it is extremely important to select the proper personal computer to suit the business applications involved, and therefore the correct operating system, storage capacity, programming language, etc. must be thoroughly investigated. We are concerned with the data communications capability of personal computers, because there are often advantages to being able to communicate with other personal computers or a large business computer.

Many personal computers offer asynchronous start-stop type communications with the associated limitations as a communications protocol. Others offer 3270-type synchronous communications with the associated advantages. Some permit the personal computer to operate as a bisynch 3780-type RJE station. Some even provide all three of those capabilities. If the personal computer is installed in a company that also has larger computers, it might be advantageous to select a personal computer that can access the larger computers.

Point-of-Sale Terminals

The two general categories of point-of-sale (POS) terminals are those designed for retail stores (department stores) and those designed for supermarkets. We shall focus on the latter while keeping in mind that the two uses are basically the same. The concept behind the term point-of-sale is that the transaction of selling an item and recording the sale information takes place right at the time the customer purchases the item. All the information concerning the sale is immediately recorded on the in-store computer disk file. Figure 7-11 shows a typical POS terminal.

A POS terminal consists of the following components:

- A light beam (laser) scanning device to read the product identification symbols which manufacturers print on the product label.
- A customer display device to show the product identification and price of each item as it is rung up.

- A register tape printing device to produce the normal cash register receipt for the customer's order.
- A scale on which to weigh a bulk purchase and also determine the specific dollar amount of the transaction.
- A manual data entry keyboard to handle the pricing and identification of any items that cannot be scanned, make corrections, etc.

Fig. 7-11   POS terminal (*IBM Corp.*)

All the individual checkout terminals in a store are connected to a control unit which is actually an in-store computer with duplicate central processing units, memory units, power supplies, and disk drives. The need for the backup of equipment components is that a breakdown of the scanning system cannot be tolerated, because it would put the store out of business. The store's control unit computer is attached to a leased or dial-up communications line that operates synchronously at 2400 to 9600 bps.

Hand-Held Terminals

Hand-held terminals serve the very special purpose of permitting the user to be mobile during the operation of entering data into the terminal. One typical application is to record the inventory of products in a warehouse; another is to read electric and gas meters. Each involves the capability of moving around and then entering data when needed. A very popular application is the recording of an order for more product in a supermarket. Figure 7-12 shows a photograph of a typical device.

In a supermarket application the user need only walk around the store and check the numbers of cases of an item needed at each shelf

Fig. 7-12 Hand-held terminal (*MCI Corp.*)

location. The item number can be obtained by moving an electronic wand over a preprinted bar-coded shelf label, or it can be entered manually via the keyboard. Either way, the quantity required is keyed in manually via the small keyboard. The terminal transmits asynchronously at up to 1200 bps either by using an acoustical coupler or by direct connection to the store's phone line. The modem is usually built into the small terminal.

8

Host Computer and Communications

Microprocessor Technology

A microprocessor could be defined as a single large-scale integration (LSI) integrated circuit or chip that performs the logic functions of a central processing unit (CPU). It contains silicon integrated transistor circuits constructed into ceramic blocks smaller than a standard domino, as shown in Fig. 8-1a,b. The semiconductor technology used to build the integrated circuits (ICs) are metal oxide semiconductor (MOS) or bipolar TTL (transistor-to-transistor logic). All tasks must be performed in a sequential order rather than in parallel as with random logic. Some microprocessor chips can be easily altered by a manufacturer or a user and so are said to be microprogrammable.

A good example of the use of a microprocessor is in CRT display and teletype or printer type terminals, which often employ eight-bit microprocessors. The microprocessor is placed between the data entry point, often a keyboard, and the printer or display. This allows the microprocessor to interface with the operation on a character-by-character basis. Through the use of random-access memory (RAM) the terminal can store pages of information which can be either recalled by the operator for review or editing or transmitted in blocks to and from another terminal or a large computer. The microprocessor allows the operator to make additions to or changes in memory rather than on hard copy and permits storage of forms, addresses, etc.

In most applications, one microprocessor, or CPU chip, is used along with a number of memory and support chips. Semiconductor memory chips are used for temporary storage (RAM) of data and permanent storage (ROM) of programs and other constant information. The ROM (read-only memory) can be programmed by the manufacturer or user; the PROM (programmable read-only memory) is a memory chip that can be reprogrammed for new applications. Semiconductor memory consists of storage elements arranged in a series of words that contain a fixed number of bits, usually 8, 16, or 32. Each word, which is accessed by an address, can store binary information. An eight-bit word, called a *byte,* can store 2^8, or 256, different combina-

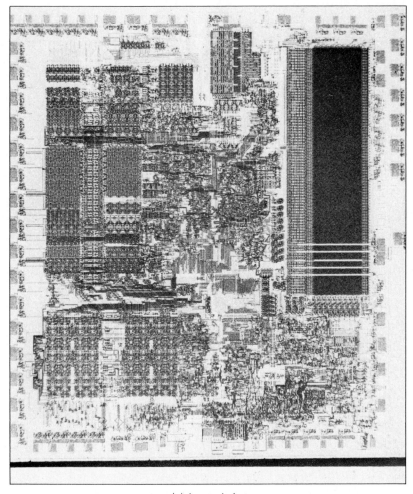

(a) *Internal view*

Fig. 8-1 Microprocessor chip (a, *National Semiconductor; b, Rockwell International Corp.*)

(*b*) *External view*
Fig. 8-1 (continued)

tions. Each of these different combinations can be associated with a distinct data value or computer function. Microprocessor and memory chips are closely associated in that the microprocessor performs the basic program operation contained in the ROM while using the RAM for temporary storage of data.

The Motorola MC68020, for example, is typical of the new generation 32-bit microprocessors that accept, process, store and release data in 32-bit increments. The MC68020 uses CMOS (complementary metal-oxide semiconductor) technology and contains approximately 200,000 transistors installed in a 114-pin package. By using a 16.16-MHz clock rate, this microprocessor can process from 2 to 3 million instructions per second. State-of-the-art RAM chips used with microprocessors are available with storage capacities ranging from a few thousand up to 264,000 bits. Typical values are 16,000 and 64,000 bits.

The logical size of a microprocessor is determined by the length of the data word with which the device operates and the number of words it can store. A 16-bit microprocessor must process 16 bits of data at a time, no more and no less. Bits not required are normally set to zero. A single bit allows only two possible combinations of information, a 0 and a 1. A two-bit word allows 2^2, or four, combinations of data: 00, 01, 10, and 11. Thus the larger words provide more combinations of data for faster and more efficient operation.

A computer is often referred to as a CPU. Actually, it is made up of three major components: a CPU, which is a microprocessor digital circuit used to interpret instructions and perform logical functions, a memory section which holds data temporarily as it is being used or changed, and an input/output (I/O) facility for communication with the outside world of disk drives, tape drives, printers, front end processors, etc. Both the memory and I/O facilities must have addresses so data can be transferred to and from those locations. The digital computer is a binary device, since only two possible conditions can exist in a circuit. The circuit must be either ON or OFF, also referred to as a 1 or 0 condition.

Memory Volatility

Memory that is lost when the source of power is removed is said to be volatile. RAM-type memory is most susceptible to the volatility problem. Often even a small power dip will destroy some of the data in the semiconductor memory, and the results of perhaps hours of computer machine and operator time will be lost. One solution to the problem is to supply an additional power system that will sustain the memory contents by sensing when the primary power starts to fail and then take over the function of supplying the power. Another solution is to utilize a different type of memory technology that retains its contents when the power is cut off. This is referred to as nonvolatile memory, and there are trade-offs to be considered in its use. A volatile RAM chip is 4 to 5 times as fast as a nonvolatile ROM chip.

There are two basic approaches to the creation of electrical conductivity in pure crystalline silicon to produce microprocessor chips. In the first approach, an n-type semiconductor is generated by replacing some of the silicon atoms with atoms containing additional electrons, such as phosphorus impurities. This provides mobile electrons that are free to move as conductors of electricity. In the second approach, a p-type semiconductor is generated by introducing atoms such as those of aluminum or gallium, each of which has one electron less than the normal five of pure crystalline silicon. This results in the creation of openings, or "holes," for the movement of electrons. The

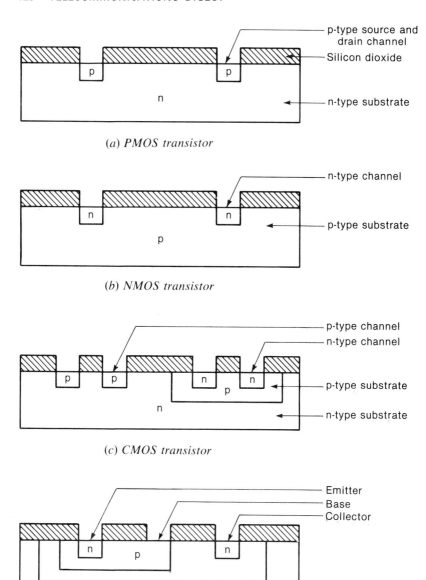

(*a*) *PMOS transistor*

(*b*) *NMOS transistor*

(*c*) *CMOS transistor*

(*d*) *Bipolar transistor*

Fig. 8-2 Types of microprocessor transistors

bipolar IC transistor wafer structure is formed by accumulating a layer of silicon oxide onto an *n*-type substrate followed by a *p*-type diffusion.

PMOS was the first LSI technology. It is a metal-oxide semiconductor that uses *p*-type channel transistors. It is easy to fabricate and affords high component density; for example, 0.25-in-square 32-bit microprocessors can contain 450,000 elements. However, it has a lower operating speed (3 MHz) and requires a negative power supply. NMOS technology for LSI components requires the reversal of the *n*- and *p*-type conductors used in PMOS conductors. Here the electron mobility is increased, since the excess electrons, rather than the holes of the *p*-type conductor, are used for conduction. With NMOS the chip size can be smaller and the operating speed faster. In CMOS technology both *p*-type and *n*-type conductors are used on a single substrate. Since only one of the transistors is turned on at any one time, the CMOS technology offers the advantage of lower power requirements. However, the CMOS device has a lower circuit density and is more difficult to fabricate (see Fig. 8-2).

All of the MOS technologies mentioned thus far suffer from the problem of volatility. They assume a random state when power is removed. A new semiconductor technology, metal–nitride oxide semiconductor (MNOS), resolves the problem of volatility and can be used where nonvolatile memory is critical. Its disadvantages are that it requires larger voltages for operation and longer times for functioning.

Microcomputers

The term microcomputer is used to describe the smaller computers. The major components of a microcomputer are shown in Fig. 8-3. The microprocessor consists of a control unit and an arithmetic and logic unit (ALU). The control unit coordinates the operation of I/O devices such as disk drives and printers, inputs raw data into the system, and sends the final result back out again. The ALU can, as instructed by the control unit, store data that has been obtained from the data memory or the I/O devices and perform basic logical and arithmetical operations such as addition, multiplication, and division. The programmed instructions that reside in memory are accessed by the control unit, interpreted, and performed. There are two categories of memory: data

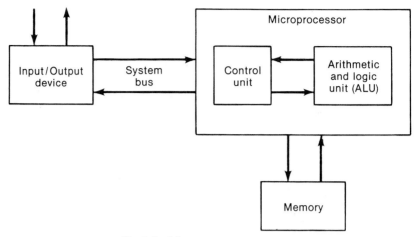

Fig. 8-3 Microcomputer components

memory and programmed instruction memory. They can reside in distinctly separate memory chips or be combined into one or a group of memory chips, depending on the microcomputer design. The various I/O peripherals are connected to the central processor by bundles of wires called the *system bus*.

Single-chip microcomputers, containing on-board RAM and ROM memories, are available for specialized applications. The required programming for the ROM can be accomplished by either the manufacturer or the user. Smart modems and display controllers are typical applications for this versatile type of microcomputer on a chip. The INTEL 8084, with 1000 bytes of ROM and 64 bytes of RAM, and the Zilog Model Z-8, with 2000 bytes of ROM and 128 bytes of RAM, are representative of available single-chip microcomputers.

Microcomputers, being smaller than mainframe computers, usually contain their own data communications line adapter hardware rather than go through a separate communications front end processor. The I/O port of a microcomputer system is the interface for data communications with terminals and other computers. An asynchronous data transmission between microcomputers A and B could take place as described below. A places a 0 in its request to send (RTS) channel. B periodically monitors the RTS channel and, upon encountering a 0, sends a 0 on its clear to send (CTS) channel. Now A can send data to B. The DATA channel is initially in a 1 state, and so A transmits a 0 start bit. Through its synchronized timing, B samples the middle of the start bit and then the middle of the subsequent data bits. During the next stop bit, A can indicate the end of the transmission or B can remove the CTS signal if more time is needed to process the data already received. The length of the transmission can either be indicated in the first data byte or be prearranged.

A full duplex operation would require two sets of RTS, CTS, and DATA channels. Other refinements such as error detection and correction, interruptable transmissions, and variable-length transmissions could be accomplished with additional signaling procedures.

Each central processor is, of course, busy performing other functions in between the data communications transactions. Rather than require that the central processor continuously take the time to check for RTS, an interrupt system could be provided to permit the central processor to store the task in progress when a communications line requires tending. The communications activity must be given a high priority or data could be lost. When a data communications function is completed, the central processor returns to its normal activity that was stored temporarily.

Mainframe Computers

A mainframe computer, as shown in Fig. 8-4, is also referred to as a host computer when data communications is involved. Most data communications networks are built around a central source of information, or even multiple sources, usually identified as a centralized database. This centralized master record file at a headquarters location can be updated periodically by, say, several decentralized shipping locations. This exchange of information between terminals and the centralized database(s) can occur in *real time*—right when the transaction takes place—or it can be performed in a *batch mode* when computer processing and communications line facilities are more readily available and perhaps more economically priced.

The terms interactive, conversational, on-line, and real time are used to describe the relations between the user (i.e., data entry operator or person using the computer system) and the central processor or mainframe computer. Reference is to an immediate exchange of information, such as inquiries and responses or entry of data and acknowl-

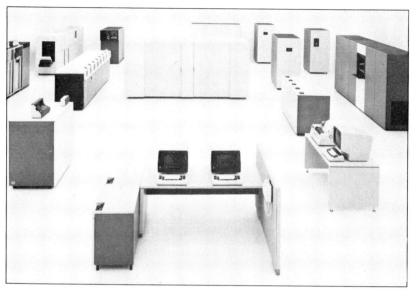

Fig. 8-4 Mainframe computer (*IBM Corp.*)

edgement. In this mode of operation the individual operator of a data entry terminal, such as a CRT display, is connected to the host computer either continuously throughout the day or for only selected periods of time as needed. Periodic access to the host computer could be managed by means of a phone call; but once the connection is made, the interactive back-and-forth communication is basically the same as if a leased line were being used.

The host computer facility can be divided into two major components: the hardware and the software. The hardware consists of the physical components that we can see, such as the microprocessor chips, disk drives, magnetic tape drives, printers, card readers, system CRT consoles, etc. The software consists of the programmed instructions that are read into the computer to make it operate and perform the functions desired. Each of the components will be covered in a little more detail below.

Hardware Components

The major element, or "heart," of a computer is the CPU, which is composed of microprocessor chips. The storing of information is normally confined to memory-type chips. The chips can be read-only memory (ROM), which consists of fixed instruction sequences containing information that cannot be changed. For smaller computers, and particularly for intelligent terminals, instruction programs are "burned" into memory as permanent programmable read-only memory (PROM). Erasable PROMS, or EPROMS, are memory chips that can be reprogrammed to meet new requirements. The bulk of the memory chips in a computer are set up as read/write memory; programmed instructions and data are read into memory for one business application and then written over later for another application. Peripheral memory such as disk and tape drives is used for the larger quantities of data and programmed instructions that are needed less frequently and need not occupy the more critical central processing memory. Disk drives are usually used for random-access memory (RAM); the data can be found immediately (randomly) when needed, read into the CPU for processing, and left unchanged, or it can be deleted or written over if it is no longer needed. Tape drives are restricted to sequential file organization in which data cannot be readily skipped over to obtain what is needed. To get to needed data that is in the center of the tape, half the tape must normally be read. Thus a message switching center would have to store its received messages on a disk file so the participants could retrieve the messages at random when the terminal facilities become available.

Mainframe computers also have channels, which consist of bus and tag cables used to connect peripheral equipment such as disk drives and front end processors. Figure 8-5 shows a typical computer system arrangement. Some of the smaller host computers have the communications line connection hardware, called line adapters, included as an integral part of the computer. For instance, the IBM 4331 computer can optionally have its own internal line adapters that are limited to attachment of up to 64,000 bps of data communications traffic lines. The larger host computers, like the IBM 3081, require that a separate front end processor, like the IBM 3705, be attached to one of its channels as a peripheral data communications line attachment device. The use of this separate front end processor (computer) greatly expands the capacity of the data network; it permits up to 352 lines, depending on the speeds and protocols used.

The actual processing capacity of a mainframe computer is rated in millions of instructions per second (MIPS). A typical mainframe computer can operate at a speed of 14 MIPS; a small microcomputer

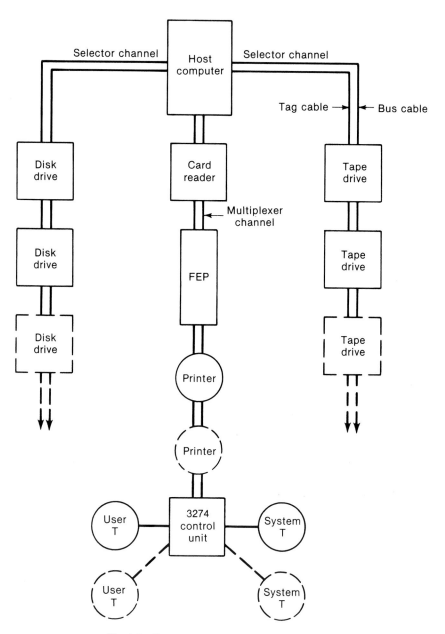

Fig. 8-5 Host computer hardware components

can operate at less than one MIPS. This speed is needed because the computer basically acts on only one instruction at a time. Thus the number of instructional machine cycles required for the computer to perform becomes important; for when the computer becomes overloaded, all users must share the inconvenience of a slowdown in operating time. That is why it becomes important for a front end processor (a separate communications computer) to take as much of the burden off the mainframe computer as possible. The central processor of the host computer services the peripherals, such as printers, disk drives, and front end processors, on the basis of *interrupts* issued when the peripheral devices require attention. Communications line adapters do not have large buffers in which to store data that can be received continuously, so they must be given one of the highest priorities in contending for the host computer's machine cycles. If the communications ports are not serviced immediately, data can be lost. With most protocols and terminal types the computer can signal the sender to stop transmitting data if the central processing unit is overloaded (e.g., X-ON and X-OFF for some asynchronous terminals). It should be noted that, in Fig. 8-5, the front end processor (FEP) is preceded only by an unbuffered card reader on the multiplexer channel, so that it gets almost first choice of all the data sent down the computer channel. Each device on the channel, of course, takes only the data that is addressed to it. When a fully functional front end processor is not available (e.g., NCP is not installed), it is not uncommon to have a situation in which 20 percent of the host computer's machine cycles are devoted to servicing the data communications lines.

Software Components

The software components consist of the programmed instructions that are entered into the host computer to make it perform the tasks that are needed by the various users of the computer. These instructions are written in a variety of programming languages ranging from machine language Basic Assembly Language (BAL) to COBOL, FORTRAN, PL1, and so on. The commonly used programming instructions that run a business application computer are listed below in order of their sequence from the central processing unit out to the operator of a data communications terminal using the system. (A diagram of the relations is shown in Fig. 8-6.)

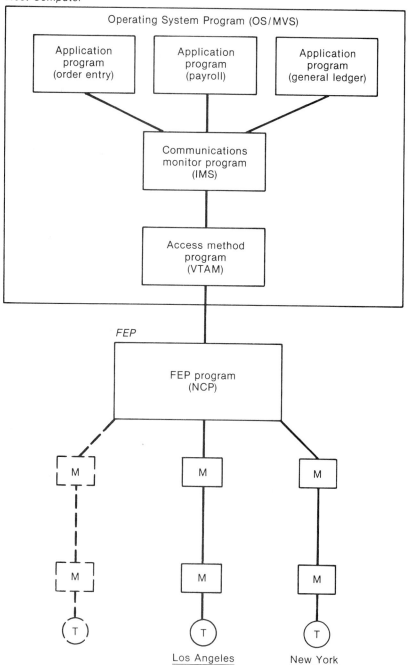

Fig. 8-6 Host computer software components

Business Application Program—The business application program is a set of instructions that are coded by a programmer according to the specifications designed by a systems analyst who is in close contact with the ultimate using departments of the company. The application programs take the form of such procedures as order entry system, accounting systems, inventory control systems, and personnel systems. These business systems are the reason for installing computers in the first place, and they are the true justification for the use of business computers. Users can inquire into the database files to obtain the details of a customer's insurance policy. They can also modify an existing policy or add a new one. The point we are making here is that it is actually the application program that defines what a user can do with a computer. Of course, other types of computer programs are needed to make certain that necessary facilities, e.g., a data communications program, are available to the user.

Operating System—An operating system is a set of programmed instructions that manage all of the available resources of a computer, including the CPU, main storage, I/O devices, and any programs that are part of the computer system. It consists of a control program with a number of optional processing programs such as language translators, utility programs, and sort-merge programs. The purpose of the control program is to efficiently schedule, initiate, and supervise the work performed by the computing system. Computer programming personnel use job control language (JCL) to direct the operating system as to the desired method of executing the business application program. Some examples of an operating system are IBM's DOS (Disk Operating System), OS/VS (Operating System, Virtual Storage), and OS/MVS (Operating System, Multiple Virtual Systems).

Unfortunately, operating systems are written for computer manufacturers to satisfy the particular needs of a specific computer or group of computers, concerning file arrangement, etc. Thus it stands to reason that the operating system of a certain microcomputer may not be compatible with the operating system of certain mainframe host computers. An attempt has been made by IBM to resolve the problem with its Virtual Machine (VM) mainframe operating system. In addition to the normal operating system function of providing an environment for development and running application programs, VM permits other operating systems to run within it. Another operating system that has become widely used and represents somewhat of an industry standard is AT&T's Unix Operating System.

Access Method—The programmed access instructions control data file organization and perform the task of moving data between the business application program and the terminals in the network. They direct the flow of data between the network elements. Examples of communications access methods are IBM's BTAM (Basic Telecommunications Access Method), TCAM (Tele-Communications Access Method), and the most recent, ACF/VTAM (Advanced Communications Function for the Virtual Telecommunications Access Method). The functions of ACF/VTAM are listed below:

- Controls the allocation of resources.
- Transfers data between points in the network.
- Permits application programs to share the network resources such as the lines, control units, and terminals.
- Establishes, controls, and terminates access to points in the network.
- Permits monitoring and altering the network.
- Detects and corrects problems in the operation of the network.
- Provides for the accessing of up to four host computers by the same terminal.
- Permits application programs and terminals in one host computer to communicate with those in other computers through the use of an associated software program called Multi-System Networking Facility (MSNF).

Another software program closely associated with the programs residing in the host computer represents the final software link to data communications terminals. It consists of the programmed instructions that run the front end processor, such as NCP for the IBM 3705 and 3725. It is discussed in more detail in Chapter 6, Front End Processors.

9

Communications Services

The ultimate accomplishment in properly resolving a communications problem is not necessarily the design and installation of a private network used only by your company. Yes, that would represent the greatest challenge and lead to the most interesting and complicated task, but it may not represent the most economical solution. In general, one could say that systems with considerable communications activity would most likely be able to support a private network, but even under high-traffic-volume situations a public network may best satisfy the needs of the business application either physically or economically. We would not want to install a private network and later have someone come along and show a dollar savings by changing to a public network. Remember that it usually takes heavy volume in concentrated areas to beat the cost benefit of a public network. Some of the most popular public network systems are described below, as are the common-carrier communications line offerings with which to construct your private network.

AT&T and the Bell System

The recently reorganized AT&T is now made up of five major subsidiaries: AT&T Communications (long-distance service), AT&T Information Systems, Western Electric, Bell Labs, and AT&T International. The Bell Operating Companies (BOCs or Bell System) with their local communications facilities have been separated from AT&T and its long-distance communications facilities. The communications line offerings of the new companies are described below.

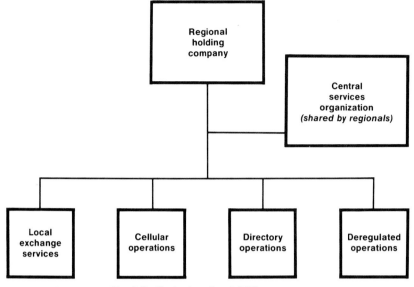

Fig. 9-2 Typical regional BOC structure

duce the newly devised Digital Transmission Service appear to be, more recently making progress. This service permits a user to make a microwave transmission to a central location which is then relayed to its destination, again via microwave. It is intended for use with wideband data transmissions, and it appears that it will soon be offered by one or more of the Bell Operating Companies. If some other company were to offer a service that "bypasses the phone company's local loops" it could represent severe problems to the economics of the Bell Operating Companies.

Western Union

Besides the telegraph service for which it was initially known throughout the United States, Western Union supplies leased lines for both voice and data and owns both the TELEX (United States and overseas) and TWX public teletype networks. An affiliate, Western Union International, provides international communications. Western Union launched the first two communications satellites, Westar, and leased channels to subscribers.

Western Union also offers leased line services in the range of service (up to 300 bps), voice (up to 9600 bps), and broadband (up to 240,000 bps). It has a special one-way transmission service for speeds up to 300 bps between certain larger cities. Companies can share transmission channels by using time-division multiplexers on Western Union's Data-Com offering. Another offering, Info-Com, permits customers to set up their own computer-controlled shared network by using Western Union's message switching computers, network control, and maintenance facilities. The company also offers a Metro-1 dial-up network between major cities, for data, voice, and facsimile, that provides reduced rates for long-distance calls.

International Telephone & Telegraph

ITT originated as the largest international communications company in the world. It provided radio and cable communications facilities overseas. Domestic carriers' lines are used to get to one of the three gateway cities (San Francisco, New York, Washington), and then ITT takes the communications overseas to the destination. ITT also offers an international Telex system compatible with Western Union's Telex in this country, and it has expanded to include a public data transmission network, called Spectrum, in the United States. It offers a long-distance network.

RCA Communications

RCA, much the same as ITT, offers international voice, data, and Telex services to foreign countries. It also has the RCA American satellite network of earth stations connected to companies across the country with leased line services for both voice and data up to 9600 bps. An associated wideband digital service runs at 56,000 bps between RCA earth stations and any customer-owned earth station.

Microwave Communications, Inc.

MCI has built a microwave network throughout the United States with the intention of establishing itself as "the long-distance telephone company" with reduced rates. It provides dial-up lines for both voice and data, plus leased narrow band and broadband lines.

General Telephone Electronics

GTE is the parent of a variety of communications companies, including the largest group of non-Bell local telephone companies. It is firmly entrenched in both the microwave and satellite communications businesses; it offers both dial-up and leased lines and launched its own satellite in 1984.

GTE offers a full line of PBX systems plus voice and data terminals for integration into its PBX system so customers can obtain text and data information such as GTE's Telemail electronic mail service via the telephone network. It offers networks with analog facilities and direct digital interface with 56,000-bps data switching. The Telenet packet switching network is another GTE company.

In 1983 GTE acquired the facilities of the previous Southern Pacific Communications Co. and formed two new separately operated companies. One is GTE Sprint Communications, which is a public microwave and satellite network that is offered as a dial-up alternative to direct distance dialing. The second company is GTE Spacenet, which offers satellite communications.

Facsimile Networks

Graphnet Systems, Inc., and ITT Domestic Transmissions Systems are two companies involved in public facsimile networks. The Graphnet network was set up as a means by which facsimile users around the country can send printed or written documents to each other at reduced cost. Computers or other terminals can also transmit data which will print out on the destination's facsimile machine. The Graphnet is basically of the store-and-forward type.

ITT Domestic Transmissions Systems offers a public facsimile network that permits communications between various previously incompatible facsimile machines. There is an option to use a leased line for high-volume users, and they also have an IN-WATS service. Users merely enter a password, destination number, and message priority via a touch-tone phone.

Local Dial-up Switching Systems

A number of small companies being formed in the larger cities consist primarily of PBX-type switches plus bulk communications contracts with the Bell System and other common carriers. Client companies lease a number of dedicated phone lines (tie lines) to connect their PBXs with the switching company's central switch. Because of the low overhead involved, the rates can be less than those of any of the other common-carrier offerings.

A prospective user must be assured that the risks of a possible overloaded switch or circuits, or poor maintenance are not too large. Of course, if there are not too many such occurrences, direct distance dialing can be used for temporary relief.

Packet Switching Networks

Telenet, Tymnet, IBM, Autonet, Uninet, and other companies offer a data packet switching network in the United States. Though not all types of terminals are accommodated, the variety that are is quite extensive. Packet switching is a means by which normally incompatible terminals and computers can communicate with each other. The networks were originally installed to handle asynchronous terminals, but they have been expanded to handle synchronous terminals and even RJE printers.

Connections can be made via dial-up to the nearest packet switching node or via a leased line if the volume is sufficient to warrant the additional cost. Of course, if a location is not in or near one of the cities served by the network, a dial-up becomes a long-distance call with the associated charges. Both Tymnet and Telenet offer electronic mail with an individual mailbox for each user.

The response time on packet switching networks is reportedly very good; only a second is required to arrive at any point on the network. A protocol converter (X.25 Packet Assembler/Disassembler, or PAD) is installed at each subscriber location to convert to X.25 or a similar protocol for the network. There is a software protocol converter package for a computer that uses the network directly. Transmission from an ASCII-type terminal to a facsimile machine also is available, along with a telegram type of service to major cities.

Satellite Communications Companies

Several companies offer satellite communications for voice, data, or video transmissions; examples are Communications Satellite Corp. (COMSAT), Satellite Business Systems (SBS, a subsidiary of IBM), International Telecommunications Satellite Consortium (INTELSAT), and American Satellite Corp. With the right application and equipment, and particularly broadband transmission, satellite communications can be a good cost-effective option. Some of the services made available by the satellite companies are described below.

Customer companies have the option of either installing an earth station on their own premises for extremely high traffic volumes, or of leasing a dedicated line to the nearest earth station. The type of service could range from voice grade (and data) to 9600 bps, to a wideband facility up to 1,500,000 bps and higher. For example, COMSAT World Systems Division offers an all-digital international business service, called Digital Express. Covering virtually all forms of telecommunications, this system can handle voice, image, and facsimile, videoconferencing, or data communications at rates of from 64 kbps to 2048 Mbps. The larger 10-m antenna dishes are usually required, but smaller 4.6-m antennas also can be used. The smaller antennas can be easily moved and are less expensive.

A 56,000-bps channel can be split into segments for simultaneous transmission of voice, data, facsimile, and video traffic by use of a digital conversion unit. Multiple customers also have the option of sharing an earth station at one of their locations, say, within a 50-mi radius, for connection via coaxial cable or microwave to the other customers. Each company could then have its own 56,000-bps channel.

Timesharing Companies

Numerous service bureaus around the country offer a timesharing type of service so smaller companies can run business systems on a larger computer. Access to these timesharing networks is usually via small teletype-like terminals with keyboards and character printers, some of which are portable and have acoustical couplers.

Telecommunications Organizations

In this section some of the major organizations in the United States and Europe that are concerned with voice and data communications are briefly described.
- The FCC is an independent federal agency that regulates telephone, telegraph, radio, television, and any other type of interstate communications in the United States. Regulation became necessary because the communications business tends to be monopolistic; it is not practical to have several different companies duplicating communications line facilities to each location in the country. Companies that furnish communications services are, as they are called, common carriers and therefore must be controlled (regulated) to protect the users of the services.
- The Deutchen Bundespost in Germany is a good example of the way communications services are handled in most foreign countries. It is a government-controlled monopoly that provides both voice and data communications.
- The General Post Office in Great Britain provides all of the telecommunications, broadcasting, and mail services for the country. It is a commercial company that is regulated by the government.

- The International Telecommunications Union in Switzerland is a worldwide communications organization with about 154 member countries. It is very influential concerning standards, particularly those promulgated by the International Telegraph and Telephone Consultive Committee (CCITT).
- The International Standards Organization (ISO) is an international association concerned with the standardization of communications procedures. ANSI is one of its reporting groups.
- The International Telecommunications Satellite Organization (INTELSAT) provides a satellite system for use by approximately 104 investors throughout the world. Similar planned organizations are IMMARSAT, ECS, and AEROSAT.
- The International Communications Association is an example of one of many public telecommunications groups in this country that are open to membership by individuals who are interested in voice and data communications. This is a national organization, and there are hundreds of local groups in the larger U.S. cities.

10
Local Area Network—
Voice and Data

A general definition of *local area network* (LAN) would concern the integration of communications lines within a single building or a company complex. A simple, typical example would be the replacement of the usual tangled mass of hundreds of individual cables connecting data terminals to a host computer by a single extended cable with frequent taps for terminal connections. Further integration of the communications lines could be achieved, if desired, by including telephones, television facilities, security systems, energy management systems, machine monitors, time reporting devices, etc.

Local Area Network Applications

The primary purpose of a LAN is to reduce the amount of cables required for multiple communications devices. Recent advances in technology have extended the sphere of operations of the LAN to include contact with remote computers and terminals far from the immediate location.

In many business environments most of the communications between computers and terminals takes place within the confines of one building or a group of closely situated buildings. This short-distance communications of data, voice, and perhaps television signals falls in the category of local area networking. One application that is becoming very popular is the sharing of printers, disk storage units, and other relatively expensive equipment by a group of personal computers that can also communicate with each other.

A new idea that is rapidly growing in popularity is the sharing of telecommunications systems by the various tenants of an office building. The advantage to the users is that there are savings in the cost of design, installation, and maintenance of a single telecommunications system as compared to the cost of each tenant providing private facilities. The shared services could include a data LAN, PBX telephone facilities, video teleconferencing, an electronic mail system, and word processing.

LAN SPEEDS

The LAN differs from regular long-distance communications in more ways than just the geographical restriction. Since the communications links are relatively short and are completely controlled by the user, higher-capacity lines such as coaxial and fiber optics cables become economically practical. Thus speeds in the megabit (Mbps), or millions of bits per second, ranges can be employed to move the information quickly on the shared communications links. It should be noted that a typical LAN line speed of 1 Mbps of serial transmissions on probably one cable is just about equal to one-tenth of the speed of a typical RAM read/write data transfer, which utilizes perhaps 16 wires for its parallel transmission.

Each of the many stations on a LAN arrangement must be serviced with as little delay as possible. This requires a good means of controlling the various transmissions of packets of data sent and received by each station in a manner that appears to be simultaneous although it is, in reality, a very fast sequential operation.

LAN CONNECTIONS

At this point we must understand the difference between computer-channel-connected data stations and remote telecommunications-connected data stations. There are data stations, such as IBM's 3272 or 3274 Model 1D cluster controllers, that connect directly to a multiplexer channel of the host computer with the advantage of having the full channel speed available and not requiring any costly modems or communications lines. All of the computer peripherals, such as disk drives and fast printers, connect directly to one of the computer channel cables. Other devices, such as the IBM 3271 or 3274 Model 1C cluster controllers, must communicate with the host computer via the front end processor, which would normally require modems or modem substitutes (modem eliminators).

The devices that can be directly connected to a computer's channel cables are very limited in number, usually to only those that are manufactured by the computer vendor or made to that vendor's specifications for a direct connection. Most other devices, and they are in the majority by far, must be connected to the computer as remote data stations whether they are located right in the computer room or many miles away from it. When planning a LAN, we should keep in mind that all data stations (terminals, cluster controllers, printers, etc.) can be connected to the host computer via the front end processor as "remote stations," but only the data stations that are specifically adaptable can be connected directly to the computer channels as "local stations." The latter, if possible, would be most desirable. With remote data stations, there is the option of port sharing. If there are, say, 50 terminals but only an average of 10 are transmitting at any one time, then only 10 communications ports are required on the front end processor.

CONNECTION ARRANGEMENTS

The major categories of equipment comprising a LAN consist of mainframe computers, minicomputers, work stations (CRT displays, etc.), and personal computers. A LAN can tie these pieces of equipment together in basically four different types of connection arrangements. The telephone instruments can be included in the network, but they generally are not. One arrangement is the star network, wherein all stations are connected to a single, central controller. A breakdown of any individual station will not affect the other stations, but, of course, all are dependent upon the central controller.

In a ring network the various stations are connected, one to the other, to form a complete ring back to the controller. Each station takes only the messages that contain that station's particular address. A malfunction of a single station could "break the ring" and render the entire network inoperable. This problem is resolved with the daisy chain network, in which each station is connected to the primary ring via a secondary loop that, if broken, will not fracture the primary ring and disturb the activities of the remaining operational stations. The fourth method of connecting the various stations is a straight-line bus to which each individual station is attached as an independent connection. In the Xerox Ethernet System a bus-type connection is utilized, whereas IBM reportedly is planning a daisy chain connection network.

PBX SWITCH VERSUS BASEBAND OR BROADBAND

There are two major competing approaches to tying a company's data stations together. One is to utilize the PBX telephone switching processor. This method is normally associated with lower data transmission capacities, usually restricted to 64,000 bps. It would not lend itself to video transmissions that require broader bands, but the capacity offered would be sufficient for most companies and it has the advantage of also including all of the telephone communications. The second major approach would be to install a data LAN, such as Ethernet, utilizing either the broadband or baseband concepts to be described shortly.

To answer the question whether to go with a PBX voice and data type of local communications network or to install a baseband or broadband data LAN would require a detailed analysis of each company's particular data, voice, and video requirements. However, it is possible to make a general statement of the significant factors and relations involved. If a particular system is heavily oriented toward voice communications and/or there are numerous well-dispersed terminals requiring access to multiple host computers and to other terminals, then a PBX-based system is likely to be the better option. On the other hand, if the system features bursty, high-speed host computer–to–host computer traffic and there are fewer, or clustered, terminals, then a baseband or broadband data-type LAN might be the better choice.

The use of the PBX switch approach permits the sharing of resources (devices and software) between voice and data transmissions, as well as the consolidation of the network management function. Since both voice and data look the same to the system, some of the advantageous voice features, like least cost routing, callbacks, conference calls, etc. can now be applied to data transmissions. The PBX data/voice switches have concentrated on the connection of numerous dissimilar terminals by providing the necessary format and protocol conversions. This leaves the user free to take advantage of the flexibility of mixing various vendors' equipment on the same network to share printers, data storage, and other peripherals. As the data-to-voice proportion of the transmission load changes (i.e., data is predicted to increase from 2 to 40 percent of the total PBX traffic over the next five years), the system need only be reconfigured on the PBX switch.

Use of Telephone Lines for Data Terminals

The use of existing telephone wiring for the transmission of data to the mainframe computer can be approached with different degrees of complexity, depending on how far it is desired to go and what functions are to be served. The simplest step would be to merely add the data terminals to the phone wiring to avoid pulling separate cables for the data terminals, which is described in the first section that follows. We could go a step further and add the second function of a data switch so the various terminals could be connected to the various computers and share ports, as described in the second section. The final step to add the third function would be to combine both the voice and data signals into a single PBX switch so both types of communications can be switched locally and outside the building, as described in the third section.

Use of Existing Phone Wiring

The easiest way to integrate communications lines within a building or complex of buildings would probably be to use the existing telephone wires that extend to every office and desk without the need to pull new cables from each desk location to the computer room. It is the least expensive type of local telephone–terminal system. It does have its limitations in that it would not accommodate excessively high volumes of wider-band communications as afforded by coaxial and optical fiber cables, but it could do a good job in most unsophisticated applications such as are found in many companies today.

The basis of the above arrangement is that the existing telephone network in a building does not utilize the full capacity of the twisted pair wiring system. Thus the unused capacity can be set up to carry data signals to each desk location by sharing the existing two-wire circuits between voice and data communications.

The coexistence of voice and data signals on the same pair of wires requires that each system utilize its specific range of signal frequencies. Voice takes the lower frequencies of, say, 300 to 3400 Hz, and data takes the higher frequencies. This necessitates the use of a frequency-division multiplexer to split a single communications line into two different frequency channels. Also, some type of modem is required to provide the EIA interface voltage signals between the data terminal and the line and also modulate the data digital signal pulses so they can travel along with the voice analog transmission. So we end up with a line driver or limited-distance type of modem that has the optional feature of frequency-division multiplexing.

Each of the remote (desk location) terminals requires a separate stand-alone multiplexer modem, and a bank of many rack-mounted multiplexer modems is needed at some location, probably in the computer room, where separate voice lines can be taken off and run to the

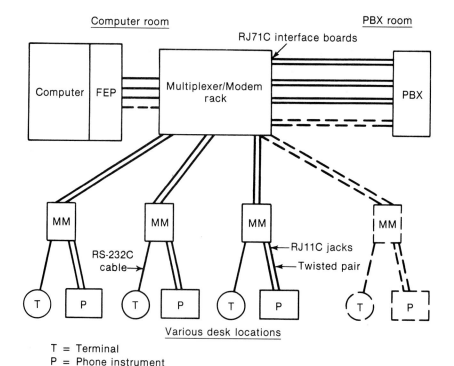

T = Terminal
P = Phone instrument
MM = Multiplexer/Modem

Fig. 10-1 Telephone line voice/data system

telephone PBX switch. A diagram of the arrangement is shown in Fig. 10-1.

The telephone line voice/data arrangement can, if properly engineered, work with an off-premise PBX and also with the multiline key telephones. Individual phone instruments are connected to the remote multiplexer modem units via a standard RJ11C modular jack, and the data terminals take the usual EIA RS-232C 25-pin connectors. The line connections at the central separation location (computer room) use RJ71C interface connection boards. Depending on the vendor selected for the system, the remote terminals and phones can be 6000 to 18,000 ft from the central separation location, both asynchronous and synchronous transmissions can be handled, and operation can be accommodated in either half or full duplex modes.

Addition of Data PBX Switch

The next step in upgrading a telephone wiring type system would be to install a computerized switch, which is called a data PBX. An example is shown in Fig. 10-2. All four of the wires normally present in telephone lines would most likely be used for this arrangement. The data PBX would provide terminals the ability to be switched between all of the various computer resources such as shared ports on the front end processors, access to multiple applications on the same computer (word processing, electronic mail, etc.), and even access to multiple computers in the same computer room or miles away. The data PBX could also contain a protocol converter so previously incompatible terminals, such as asynchronous start-stop terminals accessing a synchronous 3270 type system, could communicate with each other. This could result in a full-blown LAN system.

INTELLIGENT SWITCHING SYSTEM

An example of a data PBX would be Infotron's Intelligent Switching System, which combines data switching, local area networking, and technical control as a network management system. The intelligent switch controls a contention environment whereby a variety of terminals can contend for the communications ports on the computer's front end processor. Communications ports are relatively expensive, and when there are data stations that are not utilizing ports continuously, it

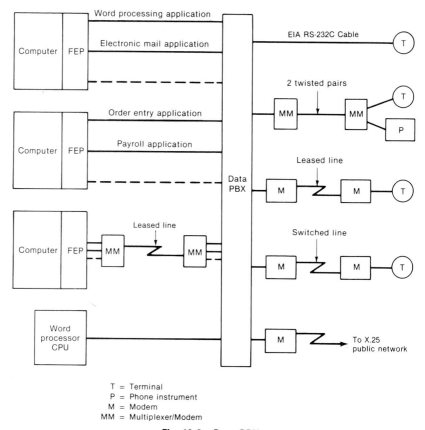

T = Terminal
P = Phone instrument
M = Modem
MM = Multiplexer/Modem

Fig. 10-2 Data PBX

is economically practical to share the available ports on a contention basis similar to a 3270 type cluster controller. The intelligent switch permits the various stations to switch between multiple applications and multiple computers. Both asynchronous and synchronous protocols are supported, the former at speeds up to 9600 bps and the latter at speeds up to 64,000 bps. The intelligent switch is also a means of bringing in a backup front end processor to replace a malfunctioning unit. The switch supports an RS-232C interface and employs token ring LAN architecture; it thereby avoids collisions or blocking delays. The nodes in the network can be up to 2000 ft apart, and there can be 64 separate nodes with 64 connections per node. The network can support 2000 simultaneous communications connections.

Addition of PBX Switch for Both Voice and Data

The final step in upgrading a telephone line type of wiring system would be to install a PBX switch to handle both voice and data in the switch itself. Of course, if the existing telephone switch is the older analog variety PBX, it will have to be replaced with the newer digital PBX so data signals can be accommodated. All digital PBX switches can handle digital data signals, but some of the newer ones are designed specifically to handle data (and voice) and contain features and data rates not possible with the earlier general PBX switches.

TIME-DIVISION MULTIPLEXING

Time-division multiplexing is used in most cases to handle both voice and data signals over the same two-wire line at each desk location. Adding the data would, of course, increase the traffic load on the company's PBX switch and may require a larger switch than would otherwise be needed. These systems were originally designed for in-house communications to a local computer, but communications with remote locations is now available.

PULSE CODE MODULATION

The most common and the only standardized method of digitizing the voice signals is by the companded pulse code modulation (PCM) technique, which converts the voice signals into a 64,000-bps digital stream based on 8000 samples of eight bits each second. Delta modulation is another method and could result in a bandwidth advantage. On the newer PBX switches the voice signals are digitized at the handset (desk location), and so both voice and data signals can travel together over the same twisted pair on a time-division multiplexing arrangement as digital signals. Thus modems are not required for the data signals.

SIMULTANEOUS VOICE AND DATA TRANSMISSIONS

An example of a digital PBX that is specifically designed to handle both voice and data is the Rolm CBX II. An individual PBX can support up to 768 bidirectional voice and data communications channels simultaneously. A complete network of multiple PBXs can be configured to up to 15 PBX nodes comprising up to 10,000 voice and data users. The total throughput of the full 15-node system is 4.4 billion bps. PBX nodes are linked together with fiber optics cables running at 295 million bps and are designed for nonblocking of signals. The voice signals are digitized at the desk locations and then both the voice and data (and some control signals) are handled on a single twisted pair of telephone wires.

CONNECTIONS TO THE PBX

The method of connecting the computer's communications channels to the PBX has undergone some changes. Originally the channel hardware circuit boards interfacing the computer and the PBX were located in a separate box attached to each communications line. In the interests of simplification and cost reduction (about 50 percent), the interface functions have been moved to the communications circuits in the computer and standard T-1 Carrier multiplexing techniques have been applied to provide 24 communications channels available with a regular four-wire twisted pair. There is a conflict, however, within the PBX and computer manufacturer group as to the specifications for this interface. There is the computer-to-PBX (or CPI) specification that has been adopted by most of the PBX manufacturers. Subsequent to that adoption, AT&T proposed its Digital Multiplexed Interface (DMI) specification that also has been adopted by some computer manufacturers. The difference in interface specifications poses a problem for computer manufacturers, who would prefer to offer only one type of interface for attachment to all types of PBXs.

The DMI interface is more packet-oriented and operates better with the future Integrated Services Digital Network (ISDN) concept. It provides 24 channels using a T-1 Carrier line at 1.544 Mbps. The CPI approach provides 23 channels using the same line. Both provide 30 channels at 2.048 Mbps when used with the CCITT standard adopted in Europe. At this point it appears that the CPI interface standard will be used by more companies for the next several years and the DMI approach can take over as the ISDN concept starts to take over.

Baseband Local Area Network

The baseband LAN method employs "baseband signaling," which means that the signals are transmitted at their original frequency without modification or superimposition on a carrier. This digital transmission approach employs time-division multiplexing, and the data is usually segregated into packets. The cabling normally consists of $\frac{1}{4}$-in

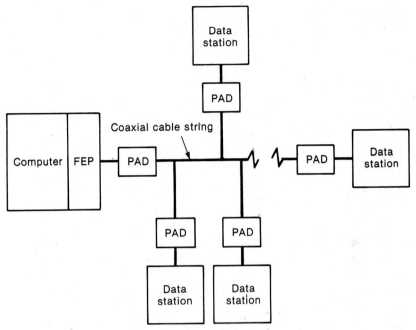

Fig. 10-3 Baseband local area network

coaxial cable that connects several data stations together in a series-type string of perhaps 1500 ft, multiple strings being used as required. Data stations are linked to the network via packet assembler/disassembler (PAD) interface boxes, as shown in Fig. 10-3. Optical fiber cable is now being offered in place of the original coaxial cable. The packets will contain the destination address, source address, control field, text data, and the frame count for error checking. The receiving station has to eventually send an acknowledgment back to the sending station. Each station PAD looks for its own address and takes only the packets so addressed. Bandwidth is an important consideration in selecting or designing a LAN for a particular company's application. Most baseband LANs operate at about 1 Mbps speed, whereas the Ethernet system runs at 10 Mbps.

LINE TRAFFIC REGULATION

One of the aspects of the baseband approach that people question is the inability to handle broadband communications such as television. Another concerns contention for the use of the single line by multiple stations. Some baseband vendors use the carrier sense multiple access with collision detection (CSMA/CD) method of regulating line traffic. It involves a statistical approach whereby each station must constantly be alert for activity prior to and also while using the line. In case of a collision, each station must back off and try again after waiting a period of time to avoid continuous collisions. Other vendors prefer a deterministic approach; they use a "token-passing frame" whereby a transmitting station is permitted a limited transmission time before it must pass the token frame to the next eligible station. This procedure, by which collisions are eliminated, is claimed to be better, especially for real-time data and voice transmissions. Token passing is a form of decentralized polling whereby each terminal is assigned a sequence number to be used in a logical ring of polling. As a terminal completes the transmission of a message, it transmits the token to the next assigned terminal in the ring. If the new terminal has no traffic to send, it transmits the token to the next assigned terminal. Although token passing is more complex to implement, the delay inherent in normal polling is reduced. Also, reliability is increased, since the system is not dependent upon a central controller.

ETHERNET

Xerox Corporation's Ethernet LAN system has been accepted as a standard by the Institute of Electrical and Electronics Engineers (IEEE), the European Computer Manufacturers, and a group of over twenty U.S. equipment vendors. Xerox offers 20 products that operate on its system, and 80 other vendors have announced plans to manufacture products that will be compatible. There were over 500 Ethernet installations at the time of writing, but the issue of the best approach to the LAN was far from settled. IBM, for instance, does not favor the Ethernet bus architecture but has endorsed a ring-network approach with token passing.

IBM's LAN

Though IBM's actual LAN system is not expected to be made available until 1986 or so, the cabling for the system has been announced so companies can get ready for it. Each office workstation will be provided with a multipurpose communications outlet that will accept a telephone jack and/or a data plug. Workstations on each floor of a building will be connected to a *wire closet* in a *star-wired* fashion. A workstation can consist of a data terminal, a telephone, or a small–to–intermediate size computer. A PBX can also be attached to the IBM cabling system. The cabling itself can be varied from two twisted pairs of telephone-type wires to fiber optics cables.

AT&T's LAN

AT&T Information Systems now offers a LAN system similar to that planned by IBM. It is called Information Systems Network (ISN), and it is based on the token ring network concept with star-based wiring, centralized control, multistation line card modules, and twisted pair wiring using four-pair, or quad, cable for each device attached. The hardware consists of three short bus-type local networks inside a central controller. One of the buses handles the transmissions, one the receiving side, and the third the contention for access to the network. All three operate at 8.64 Mbps. There is also a provision for high-speed optical fiber links that can be connected to remote concentrators up to a kilometer away from the high-speed transmit/receive bus. The packets themselves are quite unique in that they contain only 180 bits of data, and several different devices can share the same packet. Packet switching is set up either as a permanent virtual circuit, in which the terminal is always logically connected to the same destination device, or as a switched virtual circuit, in which the terminal is connected only temporarily to almost any device. The ISN system provides for 42 plug-in modules representing 336 local terminals. The total ISN can support up to 1200 virtual circuits and 1600 terminals.

Broadband Local Area Network

The term broadband is applicable to a LAN because a single cable can accommodate up to 150 Mbps, which is several times the capacity of a baseband coaxial cable. The physical broadband cable is also of the coaxial type, but it is a little larger in diameter than the baseband cable. Further, its outer cylinder conductor shield is usually a sleeve of extruded aluminum rather than a woven mesh of copper wire. It is also known as cable television (CATV) cable.

The broadband LAN system differs drastically from the baseband system in that the data packets are transmitted as analog, not digital, signals. Thus modems are required for digital-to-analog conversion. This introduction of modems that operate in the megabit range introduces one of the disadvantages of broadband LANs, namely, the increase in cost and complexity. On the other hand, the greater frequency range affords a variety of predetermined subchannels such as 48 full duplex channels at 9600 bps, plus 128 switched channels at 9600 bps, plus 32 full duplex channels at 64,000 bps, plus one high-speed channel at 10 Mbps, plus five television video channels at 6 MHz.

The broadband LAN employs the higher radio frequency (RF) transmission and thus requires special RF modems at each data station. There are two methods of sharing the single line with multiple

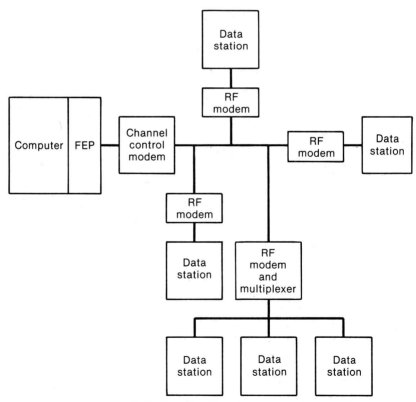

Fig. 10-4 Broadband local area network

data stations, plus other video and audio devices. A frequency range of something like 300 MHz (million cycles per second) makes the LAN an excellent candidate for frequency-division multiplexing, which is the first method used to share the line. The second method is to install time-division multiplexing. In fact, a combination of the two opens up the LAN capability to be used by a large number of individual terminals, etc.

The broadband approach offers a high signal-to-noise ratio and thus a very low bit error rate. It is very adept at handling multimode communications, mixing data, voice, television, etc., all on the same network. Just as with baseband, the broadband LAN does not accept just any data station; it is restricted to the stations specified by each particular vendor. Figure 10-4 shows a typical broadband arrangement.

11

Message and Packet Switching Networks

The message and packet switching approaches to multipoint data communications are very similar in that both accomplish about the same goals although they differ in the manner by which the actual data characters are transmitted. In a message switching system, the entire message is normally received at the central computer node and stored for a period of time, short though it may be, prior to retransmission of the entire message to the remote destination. This accounts for the name *store-and-forward communications system.* In a packet switching system (i.e., value-added network), the data to be transmitted is divided into small "packets" of data limited to something like 128 bytes (characters). The packets are usually relayed from node to node until the ultimate destination is reached, without having to be stored on disk at any one communications node. Circuit switching is another method of communications switching, where a microprocessor (computer) establishes connections between the various points on a network, but the information being transmitted is not read into the microprocessor's memory or disk as a store-and-forward function.

Message Switching Network

Message switching is a means of permitting a multitude of terminals and computers at different physical locations around the country (or world) to exchange data by using a common communications line or network. The network can be privately owned or it can be open to public use so a variety of businesses can utilize the common communications lines.

The earliest type of message switching system was in operation prior to the introduction of computers. A central location, say in Chicago, would receive a message, say from Los Angeles to New York, which was then punched into a paper tape. The tape was torn off the teletype machine connected to Los Angeles and put into the reader of the teletype machine connected to New York and transmitted to its destination. The paper tapes could be either stored until an appropriate transmission hour or retransmitted almost immediately. Thus evolved the name *store-and-forward paper tape switching system*. This function remains the same in today's message switching networks, except that the centralized operation is completely automatic and is performed by a computer instead of by a person (see Fig. 11-1).

FUNCTIONS:

The three computers that comprise the three hubs of the networks shown in Fig. 11-1 receive messages from the terminals in their respective areas as well as from each other. The message headers are analyzed to determine message destination, and, when there is a priority, to permit one message to take preference over other less-urgent messages. In some networks a certain amount of processing can take place, the minimum being the gathering of message traffic information for statistical reports.

The network detects any transmission errors and automatically retransmits any erroneous blocks of data. Headers are edited for incorrect format, invalid addresses, etc. All messages are stored on disk files as they are received, and they are subsequently transmitted to their destinations. The final transmission could be almost immediate, or it could take place hours later because of a time zone requirement, an inoperative destination terminal, or a prearranged schedule. Some messages are set up to be transmitted to alternate destinations if the addressed destinations are not available.

Extensive logs of all message traffic and the times of receipt and retransmission are kept by the computer. Records of the status of lines and terminals and their daily traffic loads also are kept. A system for billing the using terminal locations is included. Individual terminals can send a message only once, but they have the option of requesting the computers to "broadcast" the message to several destinations. In the design of the network there is an option to install more intelligent concentrators (Fig. 11-1) that permit the terminals in their domains to communicate with each other without routing all transmissions through

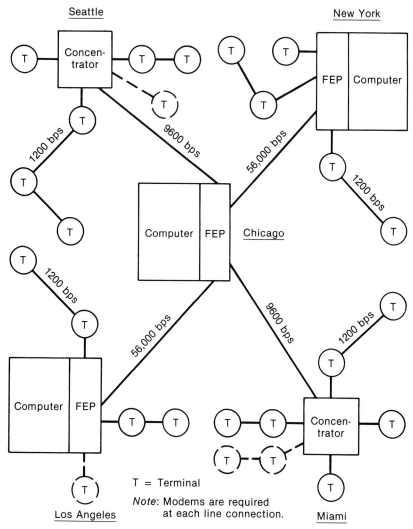

Fig. 11-1 Message switching network

a computer center. A record of each such local transmission would probably be sent to the nearest computer by the concentrator for logging and billing.

ADVANTAGES:

A message switching network can have a decided advantage over the alternative of an individual leased line to each destination from the computer if the volume of data transmitted and the spread and number of locations are great. It stands to reason that it would be easier to justify economically communications lines that are shared by many users. This advantage becomes even more prominent when public networks are involved. A private company may not be able to afford a leased line to a location thousands of miles away from any other company location, but companies with locations in the same area or along the same route could help share the cost of the line through a public network arrangement. If the network is large enough, there can be an option of alternate routing in case one of the connecting links is down, assuming that the network can support redundant or alternate communications lines. (In the example of Fig. 11-1 an alternate line could be installed to connect Los Angeles to Miami.)

Packet Switching Network

Packet switching networks are designed primarily for on-line applications in which the data must be delivered to its destination immediately. However, there can be a store-and-forward capability which in most cases is used only if a destination terminal is inoperative. Today's public packet switching networks claim a response time of a second or less within the United States. If each transmitted segment is limited to one packet of data it is termed a "datagram network." If multiple packets of data are included in a segment it is termed a "virtual circuit," which has advantages with larger messages. A "permanent virtual circuit" guarantees a continuous connection between two points while a "switched virtual circuit" provides the flexibility of a temporary connection.

A packet of data is usually 128 bytes of data, part of which includes the packet routing control information required to get the packet to its destination. A packet is sent through a communications network as an individual transmission completely independent of the rest of the sentence or block of data. The complete message is normally assembled only at its destination and not at a store-and-forward network computer node. Thus the error detection and correction function is limited to the individual packet of data rather than a block of data.

The public packet switching network operators have selected an efficient communications protocol similar to CCITT X.25 for transmission of data throughout the network, which includes hundreds of cities around the United States and even around the world. Unfortunately, most of today's terminals do not operate under the X.25 protocol. To resolve this problem, the network vendors have devised a black box, called a PAD, which acts as a combination protocol converter, a packetizer/depacketizer, and a multiplexer. Figure 11-2 shows a typical packet assembler/disassembler and data concentrator. Also available are software protocol converter packages for computers that use the network directly without a PAD. Figure 11-3 is a diagram of the basic arrangement of the network. A packet switching network can be designed with only one main switch, such as a satellite which broadcasts all packets to all stations in the network, as in the ALOHA system originated at the University of Hawaii.

FUNCTIONS:

The functions of a packet switching network include all of those of a message switching network. Today's public packet switching net-

Fig. 11-2   Packet assembler/disassembler (*Timeplex Inc.*)

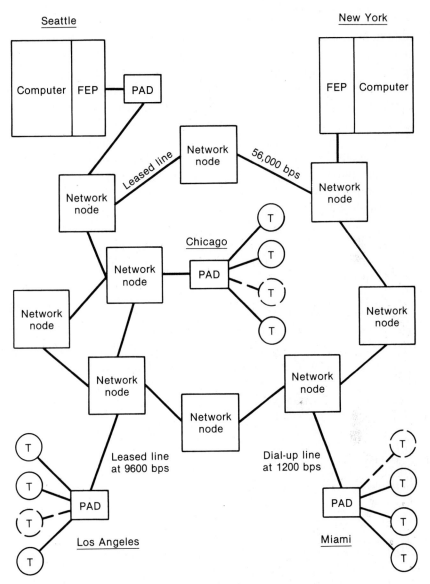

Fig. 11-3   Packet switching network

works are not looking for the store-and-forward type of operation, although they will provide a mailbox arrangement in which one location deposits a message in another location's mailbox. The second location must initiate the action to pick up its own mail, because the packet network vendors want to avoid initiating any transmissions, particularly when a dial-up facility is involved. One approach to the node-routing design is to have each sender identify all nodes on the route of a packet's destination. Since the nodes are usually computers, it is more likely that the sender will merely indicate the destination code and the node computers will find the best route for that time of day as the packet advances through the network.

Packet switching networks provide users with a terminal compatibility enhancement feature. Terminals that are not otherwise able to communicate with each other can take advantage of the protocol converter function so, for example, asynchronous terminals can talk to synchronous terminals.

ADVANTAGES:

The advantages of a packet switching network include those of the message switching network. In addition, however, there is the on-line (conversational) type of operation that affords the excellent response time that is needed for the brief transmissions between an inquiry or data entry terminal and a host computer. One of the important features of a packet network is the establishment of communications between different types of terminals that ordinarily cannot exchange data without some kind of separate protocol converter. Today's packet switching networks are also noted for their backup lines and alternate routing capability plus a very efficient, cost-effective use of the network.

Public Message and Packet Switching Networks

Tymnet

The Tymnet packet switching network offers interactive and batch data communications, but it will also provide a store-and-forward message switching function. Local access to the system is now available in over 500 cities in the United States and over 50 in other countries. At the present there are over 1200 installed nodes, 1500 host computers, and many thousands of terminals.

The network operates on the CCITT X.25 communications protocol, and any X.25 device can connect directly into the system. However, most users require packetizers/depacketizers (PADs) or a special software package that will convert a computer to X.25 protocol. The PADs, called engines, vary in capability up to one that supports 16 asynchronous channels and 8 synchronous channels simultaneously. The PADS at the Tymnet system nodes are larger microcomputer devices called Tymsats. They support X.25 devices, IBM 3270 bisynch, SDLC, and 2780 and 3780 bisynch protocols, plus a variety of asynchronous devices.

Users' terminals connect to the Tymsat nodes via dial-up or leased lines. Dial-up users outside the node's local calling area can place long-distance calls to the node or take advantage of Tymnet's IN-WATS service. Customer host computers would normally be connected to the closest node via a leased line.

One of the most interesting offerings is Tymnet's Asynch-to-3270 Service, whereby the less expensive synchronous ASCII terminals and personal computers can dial up into the network to access IBM 3270 computer applications in either the character or block transmission mode. The dial-up speed has been expanded to 2400 bps for full-screen 3270-type applications. A new X.25 type of protocol has been developed for use with personal computers; it permits connection to up to 15 channels. It allows personal computers to communicate with host computers and other personal computers.

Tymnet also offers an electronic mail system and an access connection to the worldwide Telex network. Tymnet will set up a private packet switching network if desired by a large company.

GTE Telenet

GTE's Telenet packet switching network has been in direct competition with the Tymnet system for the past several years. It is composed of many 56,000-bps links that interconnect its intelligent nodes and, in turn, connect to local network concentrators by using 9600-bps links. The local nodes are available in over 290 cities in the United

States and over 50 other countries. Though the initial offerings were for asynchronous terminals at up to 1200 bps, the network has now been extended to synchronous transmissions at 2400 to 56,000 bps. The large list of terminals accommodated includes the IBM 3270 CRT display terminals as well as the IBM 2780 RJE workstation printers. Over 100 X.25 interface products have been approved for use on the Telenet network. The multiple host facility permits terminals to access multiple applications on multiple computers.

The customer's 3270-type terminals can be clustered with a single multiplexer PAD (TP3010) that connects to the nearest Telenet node via a single leased line. Also available are software packages that will permit the user's front end processor to operate under the X.25 protocol and thus avoid the need for hardware PADs for the groups of terminals.

A recent innovation is Telenet's 3270 Dedicated Access Facility, which provides direct leased line connections between 3270-type cluster controllers or computer systems. The network itself performs all of the packetizing and routing functions that are necessary. Another service, called GTE Telenet Interface Program, permits local and remote 3270-type terminals to establish switched connections via their asynchronous host computer to any other asynchronous host application on the Telenet network. Still another service, the Micro-Com Networking Protocol, permits communications between personal computers and other users on the system. GTE Telenet will also establish private packet switching networks for larger companies that may have the need.

Net 1000

AT&T's Net 1000 system (AT&T Information Systems) has added a service bureau type of function to the traditional packet switching system. It permits its customers to write application programs that will reside in the network computer's disk storage and be called out for execution as needed for actual processing of the customer's business information, such as editing or modifying the raw data. The Net 1000 system is geared to handle a large variety of terminals, computers, and other networks, so that a single terminal can access multiple business applications and databases.

The users connect their terminals and host computers to service points in major U.S. cities, estimated to be 200 at the end of 1984. These service points contain full-blown computers that handle the data communications and data processing functions, including interfacing, conversion, data storage, program storage, and execution.

Compatibility is provided among a large number of unlike terminals through protocol and code conversion and speed matching. The system supports asynchronous contention, synchronous contention (IBM 3780), and synchronous polled (IBM 3270) types of terminals. Net 1000 provides for both interactive two-way session communications (call service) and one-way (message service) store-and-forward communications services. Messages can be edited and stored for subsequent single- or multiple-destination (broadcast) transmission.

The system provides a variety of standard application programs for the customer's use. As an example, an IBM 3270 Format Translation Program permits the use of asynchronous terminals with a host computer's IBM 3270 system. It also allows the asynchronous terminal to switch between multiple applications and multiple host computers. Another service program permits users to download information from a host computer's database to a Net 1000 service point. Other general applications, such as the Dow Jones News Retrieval Service, are being encouraged for use on the Net 1000 system.

Accunet

Another division of AT&T, AT&T Communications, offers a simpler and more standard form of packet switching network based on a 256-character packet. The customers share AT&T's extensive trunk lines and facilities with the capability of transmitting data at speeds of 4800, 9600, and 56,000 bps. The service will permit as many as 1000 data terminals to be addressed on a single data line and, depending on the speed, up to 511 different messages over a single access line. The packet service covers all geographical locations in this country and at time of writing was planned for expansion into Canada as Accunet Reserved 1.5 Service (1.544 bps).

Autonet

The Autonet Packet Switching Network is offered by Automatic Data Processing Corp., an independent computer services company.

The current network comprises 230 intelligent communication processors, over 175 host computers, and over 8700 access ports. There is a direct toll-free access at over 250 cities in the United States, and international carriers service over 50 other countries.

Autonet uses its own packet switching protocol, including automatic answer dial-up facilities for backup in the event of line or node malfunctions. The system supports most asynchronous CRT display and printer terminals at speeds up to 240 characters per second over dial-up lines. Host computers can communicate with the system through an asynchronous interface by using the X.25 protocol.

An electronic mail system, an on-line user information directory, and Auto-WATS are offered. The latter provides volume discounts for dial-up connections exceeding 50 h of activity per month. Also, direct access connection is provided for high-volume users close to one of the system nodes.

Uninet

Uninet is a packet switching network that is offered for public use by United Telecommunications, Inc. It services over 600 commercial customers with nodes in 275 cities in the United States and international connections to over 33 foreign countries. The regional nodes, which are connected by 56,000-bps digital links, supply the protocol translations needed to connect otherwise incompatible terminals and host computers.

Synchronous terminals connect directly to the local Uninet nodes, which can accommodate 32 channels—10 of which can operate as 56,000 bps and the others at 4800 or 9600 bps. Asynchronous terminals are connected to the local nodes with a special multiplexer PAD for groups of 8, 16, or 64 devices, and HDLC protocol is used. Customers can access the system via dedicated or shared dial-up ports, including the use of IN-WATS. Interactive terminals can transmit asynchronously at 110 to 1200 bps, and remote batch terminals can transmit synchronously at 2400 or 4800 bps. Personal computers, terminals, and host computers can communicate with each other via Micro-Com MNP Protocol System, which has recently been added for 10 cities in the network. Both RJE transmissions using IBM 2780 and 370 bisynch protocols also are accommodated.

RCA Cylix

The Cylix Communications Network (CCN) is based on medium- and high-speed satellite packet switching communications. It supports interactive 3270 type bisynch or SDLC protocols, plus the Burroughs Poll/Select communications discipline.

The users transmit over a 4800- or 9600-bps leased line connected to the nearest satellite earth station. The transmission is then routed via satellite to the central Cylix site in Memphis, Tennessee. From there it is retransmitted via satellite to the earth station nearest the final destination, and a leased line is used for the last leg of the trip.

The Cylix network contains 35 regional earth stations in the United States and Canada. The network claims complete redundancy of critical components so the backup facilities can take over if a primary component fails.

Integrated Services Digital Network

The ISDN is a movement currently in progress to establish a universal public network that will accommodate data, voice, facsimile, and video communications in one major network available to all businesses throughout the country and eventually the world. In the fall of 1984 the International Communications Consultive Committee on Telegraphy and Telephony (CCITT) was scheduled to start working on proposed standards and the Federal Communications Commission (FCC) had become involved by soliciting suggestions. The ISDN concept has received more attention in Europe, where communications company representatives from the United States have attended meetings. It is thought that all equipment suppliers and communications carriers will eventually coordinate their activities toward compatibility with the ISDN system.

The new network is to evolve from the telephone network, but it is to include the additional nonvoice services, with a limited set of standard multipurpose user interfaces, including functions of packet switching, circuit switching, and concatenations of both. The service will be compatible with 64,000-bps switched digital connections, and it will eventually cover lower and higher bit rate transmissions. The evolution of the system can span two decades, and it must provide for the transmission of present space-division equipment and the interworking of other ISDNs and other networks. Sufficient intelligence must be included to provide for service features, maintenance, and network management functions. The system will contain a layered functional set of protocols based on the ISO Open System Interconnection reference model and CCITT Signaling System No. 7.

Some interested parties in the Unites States believe that the evolution of the ISDNs could generate from wideband offerings of satellite carriers or the existing public data networks of the value-added carriers. They visualize multiple ISDNs in the United States alone, rather than one network for each country or the world.

Series X.400 Message Handling Standards

The CCITT has approved a recommendation for an international electronic mail network which is thought will become a standard for all future public electronic mail networks. Several companies are already in the process of developing products that will support the CCITT Message Handling Protocols. The network will permit individuals on terminals or automatic computerized data processing equipment to send and receive text messages, facsimiles, graphs, and voice and binary data in a way similar to that in which we utilize a conventional postal system. Two options of service are available. The first, the message transfer service, provides a store-and-forward type of service for transparent data transfer including both public and private message networks. A separate envelope message is used to release the main message and transmit it to its final destination. The second is an interpersonal message service, which is similar to the current mailbox-type electronic mail systems. It is more like a typical office memorandum with associated sender and receiver information affixed as part of the message.

The individual user with a terminal could send his message to an assigned user agent. From there the message would be relayed to a message transfer agent, which provides a store-and-forward delivery service between user agents. It is the responsibility of the user agent to notify (via an envelope) the message transfer agent when a message is to be released for transmission to the ultimate user. One class of service provides the capability of manipulating the message being held by a message transfer agent, and another class of service does not.

12

Data Network Design

The design of a data communications network concerns the detailed analysis of all the many factors that make up the entire network. As in the selection of almost any equipment that comprises an integrated facility, there are trade-offs of both economic and functional attributes. That is not meant to imply there are a variety of correct solutions to a network design problem. On the contrary, it is believed that there is only one correct, best arrangement for each user's situation. It is just a matter of making sure that all of the alternatives are clearly defined and properly evaluated before a final solution is settled on. Factors that must be considered in the design of a data communications network are stated and discussed below.

Batch versus Interactive Communications

Batch transmissions are those that are sent in a steady stream in the way that RJE printers receive data for hours at a time. This implies heavy usage of the communications line and perhaps a need to receive data at a speed of 9600 bps or more in order to run the printer at its top speed.

Interactive communications, on the other hand, is usually sporadic, with intermittent periods of inactivity. It implies the sending of data to a computer and the interrupted receipt of data from the computer, the longest interruption being the period in which the user is entering the data on the terminal. All this adds up to light loading of a communications link and perhaps even lower speeds. Of course, several interactive devices, such as CRT displays, could be connected to a single cluster controller so that the combination of, say, 32 terminals could well occupy a 9600-bps communications line. Interactive communications is an excellent candidate for the benefits of statistical multiplexing.

Volume of Data and Peak Load

The volume of data could be the most important single factor in the design of a communications network. Communications link capacity should be greater than the capacity needed for the peak load, but since greater bandwidths cost more money, we cannot afford to pay for much more than we will actually use. Data volumes can be estimated in a variety of ways but, at best, the estimate usually turns out to be rough. There are software packages that can be run in the host computer to give character counts for applications that are currently in operation and are being considered for data communications. More likely an estimate will have to be made in some instances by analyzing the business applications involved to determine the average number of characters sent and received for each transaction and expanding it by the numbers of such transactions each hour of the day.

The whole volume calculation procedure has to be viewed by each segment (hour or less) of the transmission period (or day). Any peak load that occurs must be smoothed out, or it will have to be used as the final channel capacity requirement. Excessive peaking might indicate a requirement for variable pricing communications links, such as dial-up lines or public networks. On cross-country communications we can take advantage of the time differences to help even out the total data load.

Response Time Requirements

We define *response time* as beginning when data is sent to a host computer (e.g., the send key on a CRT display is pressed) and ending when return data is received (e.g., the last character received on the CRT screen). The response time is made up of two primary components: the time required for communications on the line back and forth plus the time needed to process the request on the computer. If the data is out on disk, the time for retrieving it will add to the computer processing time.

There are simple devices that connect to a terminal to measure the total response time, but more sophisticated arrangements to separate the communications line and computer processing components also are available. For instance, the IBM/IMS communications monitor provides a report that gives the message enqueuing and dequeuing time by type of application.

Response time becomes critical in interactive communications because a person is waiting to receive a reply from the computer. It can not be significant in batch transmission, which can send a steady stream of data to an unattended printer. We are concerned not only with the cost of the individual's time, but even more with the person's patience. Thus a response time of 5 s or less is considered good. There are applications that go up to 20 s, but they are usually systems that are used infrequently. The communications portion of response time can be decreased by manipulating such network components as the following:

1. Increasing transmission speed
2. Decreasing the number of terminals sharing a communications channel
3. Changing from half duplex to full duplex operation
4. Giving priority to the more critical applications in the host computer, in a multiplexer, or in a message switching network

Geographical Arrangement of Locations

The locations of remote data stations with reference to the host computer, or multiple host computers in a nodal network, is normally a given condition. However, there are many ways to approach the connecting of locations together in a private network or making the decision to employ one of the many public networks available today.

The most economical network would probably be a private leased line network the percent of utilization of which is over, say, 70 percent for a full 8-h day. Even better would be use at night for data or even voice transmissions. In order to justify and properly load long-distance leased lines, we need either one location with a heavy volume of traffic or multiple locations that are clustered fairly close together or are at least situated along the path connecting two other locations. Figure 12-1 shows a cross-country leased line network in which an attempt is made to use the shortest line mileage possible while balancing the traffic load fairly evenly among the three main legs of the network.

To fully load each of the three legs, each leg would have its own modem (or DSU) and port on the front end processor in Chicago. Any two legs with light traffic, such as during an initial trial run period, could share a port to take advantage of a moderate cost savings. This, of course, has the two separate lines with perhaps twice the number of terminals contending for use of the line. That cannot be the best permanent economical arrangement. A mesh-type network, where all points are connected to each other, may be used for high volume multiple locations.

We should keep in mind that the choice of type of leased line is important. The cost increases when we go from satellite to analog to

Fig. 12-1 Leased line network

digital, but quality and reliability also increase. Be particularly cautious of the satellite line, which can be a bargain only if the hardware and application are geared for a space transmission. To install your own microwave stations requires some fairly high volumes.

The use of public networks, especially the packet switching type, should be investigated. It is difficult to cost-justify leased private lines to locations that are considerable distances away and all by themselves. Chances are that one of the public networks has an entry node nearby that is already established, is paid for, and can be accessed via a leased line extension. Some of the public packet switching networks also are very competitive with the dial-up networks.

Use of Multiplexers, Concentrators, and Protocol Converters

If there are multiple terminal devices at the same location, consider the use of a multiplexer to permit the devices to share a single line at perhaps 9600 bps. If there are multiple terminal devices at different locations clustered in the same general area, consider the use of a concentrator to permit them to share a 9600-bps cross-country line to the data center.

Data communications has evolved in an uncontrolled progression of filling the current needs for computer-associated systems. One business application at a location is installed using asynchronous communications terminal; later a synchronous terminal is installed at the same location for another business system. Accordingly, many companies find themselves with multiple data communications networks at the same locations. It is not always practical to discard the incompatible equipment, so the option to install a protocol converter should be considered. The protocol converter connects to the communications line, multipoint or point-to-point, and then converts the protocol of each of the local independent communications devices to the protocol required by the communications line or selected for a more efficient transmission. Protocol converters can also be considered for the option to use the less-expensive asynchronous terminals in a synchronous network. Figure 12-2 shows the connection of a protocol converter.

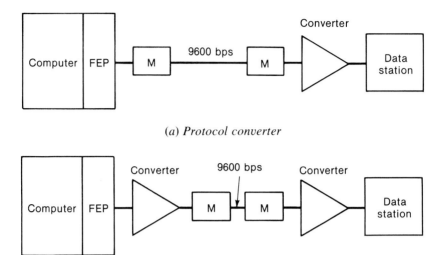

(*a*) *Protocol converter*

(*b*) *Asynchronous-to-synchronous converter*

Fig. 12-2   Protocol and asynchronous-to-synchronous converter

Note that if the computer does not handle the X.25 protocol and that is what is being used on the line, then a second converter box will be needed before the front end processor.

Asynchronous communications is normally restricted to the lower speeds of 1200 bps and less. Low-speed data terminals tend to run up line connection time costs and increase response times. With a dial-up network, a higher speed can cut down on the total connect time and thus the cost of each transmission. A relatively inexpensive way to increase the line speed is through the use of asynch-to-synch converters that provide a compatible interface to synchronous modems. Thus the data travels over the communications line synchronously at the higher speed, but the format presented to the data station or computer at the other end of the line is still the original start-stop asynchronous protocol. Some modems are available with built-in asynch-to-synch converters. Of course, if a synchronous protocol is desired at the destination end, a true protocol converter will be required at a greater expense. Figure 12-2 shows the connection of an asynch-to-synch converter as compared to that of a protocol converter.

Use of Simulation Models

There are models that can be run on a computer to simulate the selected network configuration to determine whether the network will perform as required, particularly in respect to response time. Adjustments can be made to overcome undesirable situations and rerun in the simulation model to determine the new capability. The Bell System has offered one such model for use by its customers, and similar models are available from computer vendors. The customer must supply the input data concerning locations, traffic volumes by time of day, computer processing time, protocol, type of line, etc.

Error Detection and Correction

The handling of errors can be an important factor considering the fact that a normal telephone circuit operating at 1200 bps has an average error rate of 6.6×10^{-4} (6.6 bad characters in every 10,000). The line protocol of synchronous communications performs a satisfactory job of detecting transmission errors and correcting them by retransmitting the bad blocks of data. The only concern here is that, if asynchronous communications is being considered, error detection and correction could become an important problem. Almost every asynchronous communications system can detect a faulty character and discard or identify it as such, but not many of the devices are set up to call for retransmission. A discussion with the user or the systems analyst working on the application could reveal whether identification alone is sufficient. If not, consideration could be given to the echo transmission method (send each character back to its origin for verification), the use of protocol converters, or a software editing scheme such as the use of a reverse self-check-digit for item numbers.

   A description of error detection and correction techniques is covered in Chapter 3 under Communications Traffic Protocols. The most popular method is that employed with synchronous protocols such as binary synchronous and SDLC, which are based on the ARQ principle of retransmitting any blocks of data received in error. Here the overhead consists of the 8- to 16-bit block or frame check characters, plus any retransmission required. Another technique for error detection and correction that is gaining in popularity is forward error correction (FEC). Here the bit stream is operated on with an algorithm and extra bits are added. The extra bits can be used at the receiving end to detect an error and also reconstruct any incorrect data without the need to retransmit the bit stream. Of course, the trade-off here is that the additional bits needed for the FEC approach could vary from 25 to 50 percent of the total data transmission. One advantage of the FEC technique is that the error detection and correction overhead is fixed at 25 to 50 percent. That could be exceeded by the variable overhead of ARQ techniques, in which the amount of retransmissions of data streams is dependent on the condition of the communications facilities. It would appear that designing a network in which heavy error rates are expected would involve a consideration of the FEC error detection and correction technique.

Full Duplex versus Half Duplex

Only a few years ago everyone talked about full duplex transmission but practically no one ever used it. The other two wires of a four-wire, line were used to return protocol control characters but not data. Today, especially with the newer bit synchronization protocols, full duplex operation has become a reality for many companies. It is still not just a choice of whether we want half or full duplex operation, because full duplex operation would obviously be the better approach for a leased line: the line cost increase is only about 10 percent, and the benefits are much greater. The choice is really whether we want to pay for a particular terminal device and its associated line protocol, front end processor, etc., which will provide the benefits of a full duplex operation. Even under IBM's SNA/SDLC the 3270-type terminals are still operated at half duplex, though 3705 (FEP) to 3705 and some other terminals do use full duplex. An alternative would be to obtain a full duplex operation by installing a protocol converter or multiplexers with built-in protocol converters.

The advantages of a full duplex operation become more important when we are concerned with multipoint networks and satellite communications, in which line turnarounds become a significant factor regarding time. Full duplex operation is also advantageous for point-to-point applications, as evidenced by the popularity of the

1200-bps full duplex dial-up 212A-type modem arrangements used today.

Ports on Front End Processor

A data communications network is usually controlled by a host computer that has a front end processor to handle all of the communications lines. The attachment of the RS-232C cable from the line and modem (or DSU) to the front end processor is called a port. Ports are assigned by the front end processor's software and/or hardware as being asynchronous or synchronous, half or full duplex, etc. You must make sure that the proper type of ports are available for the network you plan to install.

When multiple remote locations are dialing into the same computer for the same application(s), it is often practical to assign fewer ports than total remote locations. For instance, if 20 locations will average an hour a day of activity, spread out over the 8-h day, four ports may be sufficient on a shared basis. They can be given the phone number of the first port with a phone company "hunting arrangement" whereby a busy line will hunt to the next open line. Of course, the selection of the proper front end processor itself is an important consideration. The processor must have sufficient capacity for all current and near-future lines, and it is an important factor in off-loading data communications processing cycles from the mainframe computer.

Future Expansion Capability

We would not want to design a data network that was locked into a fixed maximum peak load, since data communications usually grows with time. Operators have a tendency to use it more as they become accustomed to its applications, and other people always want terminals too. On the other hand, to install a network 2 or more times the original traffic load for the first year or two can be economically impractical.

A network should be designed to handle the peak load and perhaps an average load of no more than 70 percent of the total capacity, if possible. Future expansion of traffic is usually unpredictable. One approach to this problem is to arrange the original network so that the expansion of the facility can be accomplished by adding resources as needed and within a reasonable lead time period.

Extra ports of the proper type should be available or be easily procured for the front end processor. Any long-lead-time items such, perhaps, as CRT displays and printers should be either on order or lined up for easy availability. Multiplexers, modems, etc., should be reviewed to be certain a reasonable lead time can be obtained. Communications links, particularly digital lines, have a long lead time, which should be considered. It could be practical to initially install a leased line to operate at, say, 2400 bps and, as the traffic volume increases, increase the speed of the line to 4800 bps and then later to 9600 bps.

Backup Considerations

The criticalness of a communications link can be of major concern. In some cases, if a line goes down, we can wait to get it fixed, but very often the business cannot afford the downtime, which could be hours. One possibility is the use of spare equipment. If you have one modem or terminal at a remote location, it probably would not be practical to pay for a spare device, but if you have, say, 20 devices, the cost of a spare may well be justified. The same thing could be said of short-distance leased lines and dial-up phones. Spare ports on the front end processor should always be available for short-term expansion and troubleshooting, and a backup front end processor will be a plus if it can be justified.

If the communications line itself is the problem, the dial-up line affords the advantage of a new connection on a redial (unless the problem is the local loop to the central office). With a leased line, *dial backup* is a very popular approach. For a four-wire leased line backup we need at each end two telephones plus an interface box that contains the switch and data access arrangement (DAA). The original modems must be capable of providing alternate dial backup. For a very critical leased line, a spare line can be kept available, or used to handle half the traffic load, and a line control facility can be made available at the data center to automatically switch to the good line almost instantaneously.

Selection of Terminals

The business application should be studied in detail to write specifications for the terminal that will provide the best overall performance. The process should be worked out with the systems analyst assigned to the system being considered. Once the basic requirements have been established, vendors that appear to meet the requirements can be contacted. It is important to talk to multiple vendors for each item and obtain quotations, including any warranties and, particularly, the details concerning service. Inquire as to the location of the nearest service center, the hours available at what cost, and the number of service people.

One good source of information concerning terminals is the current or recent user of the equipment. An opinion on performance or maintenance from an actual user is certainly of considerable value. That is particularly true when character and line printers are being considered. Printers contain a lot of mechanical parts and are notorious for frequent breakdowns, and past performance is about the only criterion for evaluation.

There are still terminals on the market that do not have buffers for the storage of characters prior to a block transmission. The transmission of individual characters can still serve a worthwhile purpose in some situations of low-volume activity, but such situations are dwindling in number. Normally, unbuffered terminals are considered less desirable.

Terminal Connection Considerations

If a single terminal is to be connected to a communications line, it becomes a simple matter of connecting a two- or four-wire telephone line to a modem (or DSU for a digital line) and then connecting the terminal's EIA cable to the modem. Most locations, however, require multiple terminals, and a single control unit (cluster controller) handles all of the traffic between the single communications line and the several terminals. A very popular example of this situation is the IBM 3270 system, which has been duplicated by many other equipment vendors.

The control unit, IBM 3274, is of two different types. One connects directly to the host computer's channel and is called a local attachment. The other attaches to the IBM 3705 front end processor via a data communications line. There are trade-offs here in that the locally attached unit affords a much better response time (the time from when a terminal operator sends a request to the host computer to when the response is received back again), often reducing it from 3 s to 1 or 2 s. On the other hand, a terminal connected to a local control unit must be within a specified distance, about 5000 ft in most cases. This distance can be doubled by the use of a special multiplexer often called a *coaxial cable eliminator* (see Fig. 12-3). A locally attached control unit will also require a host computer channel address for each of the terminals attached.

The only time the choice between local and remote attachment of a control unit becomes significant is when the terminals are within 5000 ft or so of the host computer, as in a large office building. The general rule is to use a locally attached control unit for terminals within 5000 ft of the host computer and a remote or communications line attached

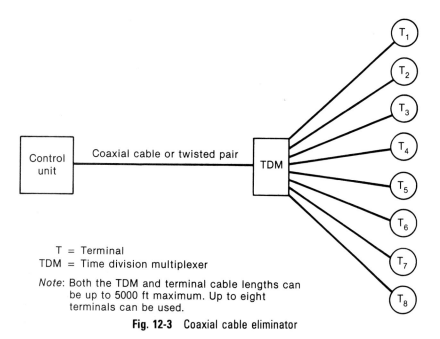

T = Terminal
TDM = Time division multiplexer

Note: Both the TDM and terminal cable lengths can be up to 5000 ft maximum. Up to eight terminals can be used.

Fig. 12-3 Coaxial cable eliminator

control unit for terminals that are farther away. If a remotely attached control unit is to be used in the same building as the host computer, it should be located as close to the associated terminals as possible to cut down on the length of the coaxial cables that connect the terminals to the control unit.

Most 3270-type terminals are connected to the control units by using a coaxial cable that has a diameter of about $\frac{1}{4}$ in. (RG 62A specification). When a large building with hundreds of terminals is wired, consideration should be given to other possible approaches in order to eliminate the bulk of the standard coaxial cables that can fill up the risers and cable ducts in a building. One option would be to use the newer ribbon coaxial cable, which comes in strips of up to 25 actual coaxial-type cables that only have about $\frac{1}{8}$ in. diameter each. Twisted pair telephone line wire also can be used, given proper engineering. Then there are the coaxial eliminator boxes (Fig. 12-3) that require only one coaxial cable (or twisted pair) to the box, which usually accommodates about eight terminals in a time-division multiplexing arrangement.

Microcomputer Attachment Considerations

The need to connect small microcomputers to a company's communications network became a common consideration with the advent of personal computers and the proliferation of all sorts of microcomputers that may or may not provide for the attachment of individual multiple-display terminals and small character printers. The network designer must decide exactly how to interface the microcomputer to a host computer network.

Most likely the network will be required to support multiple communications protocols. Very often an asynchronous interface is needed for timesharing terminals used for computer program development and other interactive applications. Of course, the network will also have to handle the synchronous communications, such as SNA/SDLC and 3270 bisynch for display screen–oriented applications. In addition, an X.25 protocol interface may be desirable to optimize the number of ports on the front end processor and provide access to other public and international data transmission networks. Another factor to consider is the capability of all terminals to access multiple business system applications in multiple host computers in the network. Centralized databases must be available to all terminal users, but at the same time, the microcomputer will want to retain control over its own local processing and retain its own local database to reduce the transmission load. Finally, all of the above must be accomplished with a maximum response time of perhaps 5 s for the processing of something less than a full 1920-character display screen input.

PUBLIC VERSUS PRIVATE NETWORK

In the design of a network to meet the above-described requirements one of the first decisions concerns the question whether to go with a public network, install a company-owned private network, or employ a combination of the two. That will call for an analysis of the company's entire data communications requirements while taking into consideration terminals, locations, volumes of data, response time specifications, etc., as described in other portions of this chapter. If a company is using IBM's SNA (Systems Network Architecture) and microcomputers, such as an IBM Personal Computer, the following general approaches to the method of attaching the microcomputers should be considered.

The simplest way to attach personal computers to a company's private network is to use private lines and provide each microcomputer with 3270/SDLC emulation software so it can appear to the network as either a terminal (physical unit, PU type 1) or a cluster controller (PU type 2). The trade-off here is that certain functions, such as the support of printers, the clustering of several terminals, and IBM's extended terminal features (e.g., different screen attributes for characters in the same field of data) may have to be sacrificed. The microcomputer can access multiple applications on multiple host computers (PU type 5) through the front end processor (PU type 4). (If a microcomputer must access non-SNA applications, it may be necessary to provide a connection to a public network that contains the interface facilities needed to cross over to a different protocol system.)

Another approach to attaching microcomputers would be to utilize the services of a public network. In addition to providing the routing, error detection and correction, transmission statistics, and associated functions, the public network would handle the protocol conversions so that the microcomputer could operate as a synchronous 3270/SDLC device. It appears that the latter function is normally offered only by public networks at the larger city locations, which could present a problem if out-of-the-way locations were involved.

Still another approach would be to install software in the microcomputer to permit it to transmit asynchronously, probably in the block mode. The microcomputer would still emulate the 3270 set of functions. Since there would be no interactive dialog between the network access node and the microcomputer, a half duplex operation at 1200 bps could result in a fairly good performance level of utilization of the available bandwidth. Block sizes could be limited to partial screens of data, so the transmission could be in progress while the user was entering the remaining portion of the display screen. This could reduce operational delays and idle bandwidth. With an intelligent microprocessor used as a terminal at one end and an intelligent host computer at the other end, the amount of data to be transmitted could be reduced to a minimum. This method would result in very low data transmission costs and, more often than not, an acceptable level of performance.

The last approach would be to connect the microcomputers to the SNA network as 3270 cluster controller emulation devices via dial-up asynchronous ports. This would require fairly sophisticated software interface packages in both the microcomputers and their associated network access nodes. The public network could supply an asynchronous link to multiple host computers of various types, other remote SNA nodes, timesharing services, and other data networks via an X.25 gate. At that stage of sophistication the network is no longer transparent to the microcomputer, because the network has been extended to include the keyboard of the microcomputer display. The use of dial-up connections permits the expansion of the network to almost any location at a very low cost.

Communications Monitor Program

The communications monitor (CICS, IMS/DC, TSO, etc.) software in the mainframe computer is very often a given condition, but occasionally there is an opportunity to select a monitor. The selection requires a detailed study of present and future communications requirements and verification that the monitor can effectively meet the requirements. Little things like the lack of handling auto-answer for dial-up ports can create severe problems. A monitor can be modified by a trained communications programmer, but modification should be avoided when possible. Needless to say, the communications monitor is an important factor in a data communications network and should be given its share of consideration.

Cost Justification Study

Our major concern should be that the network we design and install serves the purpose for which it was intended, but we also have to be concerned with the cost. Not only are a variety of configurations possible for a given network, but innumerable vendors offer equipment that varies widely in sophistication and price.

A cost justification study for a proposed data communications network cannot be made in too much detail. What always seems to happen is that the analyst fails to go far enough. He or she didn't check all of the alternatives or didn't talk to enough vendors to get new ideas on the approach. It is important that any proposed network be completely priced, including the cost of backup facilities and maintenance. If the network is considered a trial to be expanded later or dropped, approach all of your commitments on the basis of a short-term cancellation option.

13

Data Network Application Examples

Telecommunications Analysis for Information Systems

The purpose of this chapter is to present to the reader a typical, actual data communications network for a medium-sized company in the supermarket business. The point we are pressing here is that the telecommunications analyst is providing a service to the systems analyst who knows the ultimate user's needs and designs the system accordingly.

The communications network serves the business world in that it moves information to distant locations where it is needed. The information itself is most often stored in a centralized host computer and exchanged between outlying locations usually throughout the business day. All of this activity is usually provided and controlled by an information systems department representing the corporate office, or each division, if decentralized, could have its own host computer and information systems department.

Requests for data communications facilities usually come from the computer systems analyst, who has designed a business system that requires using department personnel to enter data or request it from the host computer. The systems analyst must provide the telecommunication analyst with a list of requirements including such things as locations involved, maximum response time (usually 5 s or less for interactive systems), volume of data, frequency of transmissions and time of day, and any other details that affect the communications arrangement. We cannot overemphasize that the telecommunications analyst should have a good knowledge of the business system involved in order to set up a proper telecommunications network.

Order Entry System Network

The order entry system network is a computerized system that actually runs the daily operation of the supermarket business. Twice a day, at prearranged times, the host computer calls each of the 200 stores to pick up their orders to fill the shelves with merchandise that is needed. The mainframe computer has a calling list containing the telephone number of each of the stores. The only additional hardware needed is the automatic calling unit, which converts the digital phone numbers to actual dial tones required by the dial-up telephone network.

Meanwhile, the clerks at the store have completed their task of going around the store to each shelf location and entering on a hand-held terminal the six-digit product code (or scanning a shelf label for a twelve-digit Universal Product Code) and the quantity. The store terminal is then set up for automatic pickup by the host computer during the scheduled time interval. If the terminal at the store is not ready at the scheduled time, the host computer will recycle the calling list and call the store again. The order entry system network is shown in Fig. 13-1.

The transmission from the stores is at 1200 bps and is asynchronous as dictated by the relatively inexpensive store data transmitting terminals. Since each transmission takes about 5 min, we see that two calls a day times 200 stores requires about 16 h of solid transmission. Though the supermarket workday is greater than 8 h, we can see that two communications lines are required if a perfectly scheduled and perfectly operated store calling system can be assumed. Considering the above and the fact that an asynchronous transmission has only about 75 percent throughput efficiency, we elect to install four lines. Dial-up lines are relatively inexpensive, and a port on a front end processor can cost only about $50 to $100 per month.

ORDER FILLING

The host computer writes each store's ordering information onto a composite disk file until all of the stores for a particular *picking batch* have been recorded. The order information is then processed on the host computer to create product picking label data plus store billing data, etc. The product picking labels are needed as soon as possible at the four different warehouse locations; so as soon as the batch of order data is processed on the host computer, the product picking labels are

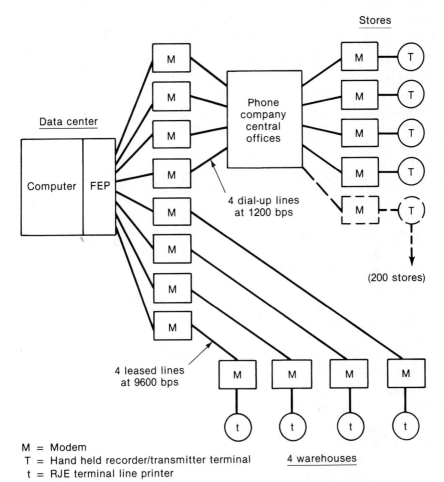

M = Modem
T = Hand held recorder/transmitter terminal
t = RJE terminal line printer

Fig. 13-1   Order entry system network

spooled out to a disk file in a 132-column print format. When the four warehouses are ready to print their picking labels, and this is usually in accordance with a close computer running schedule, the warehouses transmit the job control language (JCL) request data to run a job on the host computer that will print the picking labels on the remote job entry (RJE) printer terminal. The printers are of the fairly fast, high-volume line type that run at about 550 lines per minute on a regular No. 3002 analog leased line at 9600 bps.

The warehouses then pick the product ordered by each store by merely pulling each label off the label-holding sheet and affixing the label to the correct product case. Of course, the labels are on strips of paper in the sequence corresponding to the location of the cases of product in each picking aisle. The warehouses also transmit information back to the host computer concerning items that are out of stock and cannot be provided as ordered by the stores so the items can be deleted from the store billing. This is done with a small, hand-held terminal of the same kind as used in the stores.

Point-of-Sale System Network

There are two general categories of POS system, one being applicable to the supermarket or food business and the other to the retail stores (department stores, etc.). We will discuss the supermarket POS system, which is characterized by its numerous items on a single customer's transaction.

The POS system provides a means of scanning a symbol-printed item in a store's checkout lane to automatically ring up the sale while also obtaining product movement and sales information. The various food items must be identified by their respective Universal Product Code symbols, which are printed on the items by the manufacturers. A file of all items to be scanned must be created at the host computer centralized location for constant maintenance by the various merchandising personnel. Each store must be transmitted pricing and description information for its own particular items. There is a computer at each store that controls the operation of the up to twenty or so electronic cash registers, one in each checkout lane. The computer's disk file contains all of that store's current item pricing information. When an item is moved over the laser beam in the checkout counter, the price and description are sent to that register and the sale is recorded on the store computer's disk file. The individual store registers are connected in a loop arrangement with the capability of bypassing a defective register so as not to break the loop connection.

The data communications for this system is strictly a batch-type operation. Sometime during the evening the item file update information is transmitted to the stores. Each store can receive item file update information about three times a week for a period of about 15 min. However, once a week there is a need to pull each store's complete item file back to the host computer for verification of correctness and

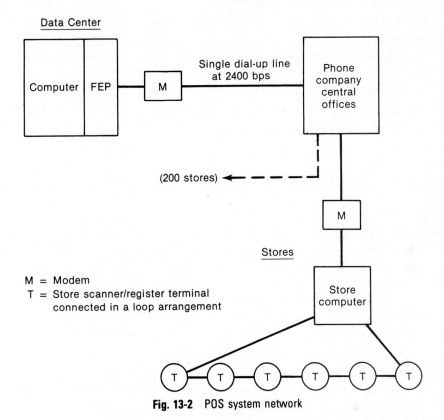

Fig. 13-2 POS system network

gathering of the week's sales information. This transmission takes about an hour for each store. Since the volume of data is of some magnitude, a bisynchronous protocol with its error detection and correction capability is required. A transmission speed of 2400 bps is selected by the scanning system vendor because it can handle the volume without spending twice as much money on a faster modem. (Of course, the scanning equipment vendor could have considered the advantages of smaller dial-up connection costs and less computer transmission time by going to a higher speed of 9600 bps.) The POS system network is shown in Fig. 13-2.

Check Cashing System Network

The check cashing system differs greatly from the other supermarket store systems in that it is an entirely on-line system that requires that the computer transaction be completed as soon as possible while the customer is waiting to cash a check. The customer's check cashing approval information must be retained in the host computer file with access provided to several automatic check verification terminals at each of the 200 stores. Fortunately, the information transmitted to the

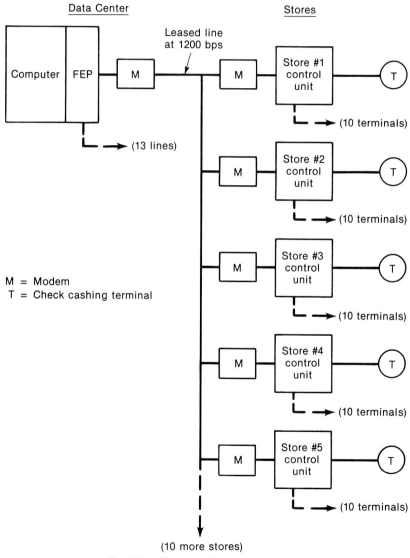

Fig. 13-3 Check cashing system network

host computer is minimal. It consists of the customer's checking identification number read directly off a plastic magnetically impregnated card, plus the amount and a secret number that only the customer knows and enters on a keyboard pad provided. With only about 30 characters being sent to the host computer and a yes or no code returning, it is possible to provide about 3-s response time while running asynchronously at 1200 bps.

All the five to ten check verification terminals at a single store are connected to a single control unit that is connected to a leased communications line. It would be impractical in this case to use dial-up lines primarily because a dial-up connection could take up to 30 s and therefore would not meet the short response time requirement. Another reason is that, even with the small amount of data being transmitted per transaction, the 2000 or so terminals would cost more to operate on a dial-up line on which there is a charge for each call. The check cashing system network is shown in Fig. 13-3. It is a leased line multidrop network consisting of 13 lines, each of which is connected to the host computer and contains about 15 stores as communications drops. This multidrop arrangement requires that the individual store units be polled by the host computer constantly to pick up any check cashing requests that are pending at any of the stores. Error detection and correction is not vital to this system; for if the host computer receives any bad data, it sends a retransmit request to the store terminal, which is holding the 30 or so original request characters in its buffer. If the store does not receive an answer within a minute or so it merely retransmits the request to the host computer.

Message System Network

The message system consists of a single teletype-like terminal in each store for the purpose of transmitting various types of information to the stores, such as incorrectly published ads, bad-check passers, and special store bulletins. This system eliminates the need for clerks at headquarters to call all of the stores every time there is a critical situation requiring action by the stores.

The network is set up as an asynchronous transmission at 1200 bps. Most of the urgent messages are short and can be handled fairly quickly on the four lines by the computer utilizing automatic calling units to dial the stores. The longer bulletins and less urgent messages are transmitted at night. The same dial-up phone lines and modems that are used for the order entry system are used for these message terminals also. Since the message terminals and the order entry terminals are relatively inexpensive, nonaddressable terminals (i.e., the computer cannot select one or the other by an address code), the host computer sends its data to whichever terminal happens to be on the communications line at the time. Since the store ordering terminal is used only twice a day, the host computer must be aware of the two half hour or so non-message-transmitting periods, which are controlled by a manual switch at the store. Transmission to the wrong terminal is avoided by the host computer's check of the terminal identification

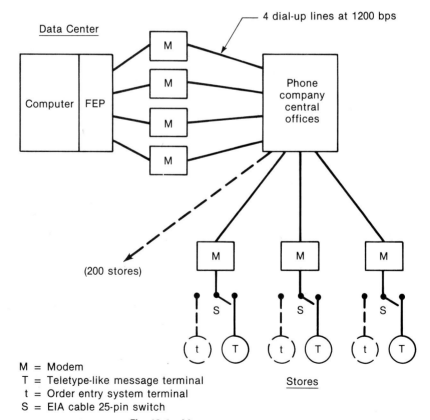

M = Modem
T = Teletype-like message terminal
t = Order entry system terminal
S = EIA cable 25-pin switch

Fig. 13-4 Message system network

number (terminal ID) prior to any transmission. The message system network is shown in Fig. 13-4.

Composite Network of Four Systems

The four different network systems were installed independently of each other over a period of years to satisfy the needs of the four different specific business systems. At the time they were installed, mixing different types of terminals on the same communications line was not possible. The following description of the four different systems points out how poorly matched the systems are for the purpose of combining them into a single multidrop leased line network. Even the two synchronous batch systems present a problem in that the inexpensive terminals are not addressable and cannot be specifically selected by the host.

> Order Entry System—Asynchronous, 1200-bps, batch operation
> POS System—Bisynchronous, 2400-bps, batch operation
> Check Cashing System—Asynchronous, 1200-bps, on-line operation with short response time
> Message System—Asynchronous, 1200-bps, batch operation

It is possible to combine unlike terminals on the same communications line by using at least two different general approaches. One is to select terminals that are supported by the same line protocol, even though they are distinctly different types. IBM's SDLC protocol under SNA will support various terminals on the same communications line, but the system designer is restricted to certain terminals specified by IBM. IBM does provide a POS system terminal, and terminals that will function under SDLC to provide the other three applications may be available. This approach presents some major problems to resolve, and it does tie us to SNA/SDLC and probably IBM terminals or a special interface to handle non-IBM terminals.

The other approach, which leaves us free to select or use our existing terminals, is to install, in each store, a protocol converter that will accept the input of the four different systems and convert it to a common protocol, like X.25, to be used efficiently on the communications network. There are protocol converters with about four input

Fig. 13-5 Protocol converter (*Black Box Corp.*)

ports and one output port which also have the capability of providing a priority to the most time-critical input line; the check cashing system in this case. It should be noted that now the order entry and message terminals can receive data at the same time, since the protocol converter handles them as if they were truly individually addressable devices (see Fig. 13-5).

A larger protocol converter containing a port for each of the multidropped leased lines will have to be installed at the host computer location. The composite network of four systems is shown in Fig. 13-6.

IMS/TSO Network

IBM's Information Management System (IMS) and Time Sharing Option (TSO) are two computer software systems that are used by most major businesses today. IMS provides a database management and data communications facility that lends itself to all types of business applications such as vendor payments, order entry, inventory control, and accounting systems. The system is designed and maintained by data processing personnel for use by all of the other "production personnel."

TSO is used by data processing people to create and maintain computer programs and associated files and can also be used by production people for business applications. TSO utilizes a data communi-

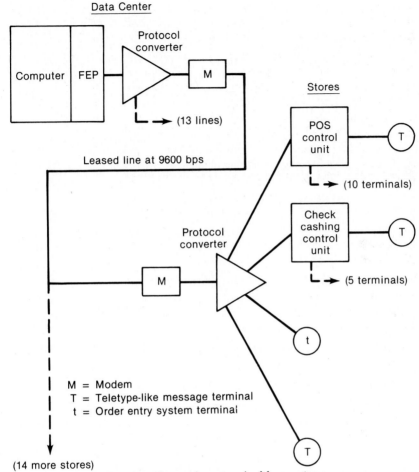

Fig. 13-6 Composite network of four systems

cations software facility called TCAM. Though both asynchronous terminals and synchronous terminals (3270-type device) are in popular use for TSO, the IMS system is normally restricted to 3270-type terminals or asynchronous terminals that are protocol-converted to operate like 3270-type devices. We will look at a 3270-type network designed for both IMS and TSO users at the corporate office and a large divisional remote location as shown in Fig. 13-7.

At the remote location there is a total of 120 terminals, five of which are character printers and the rest CRT displays for data entry.

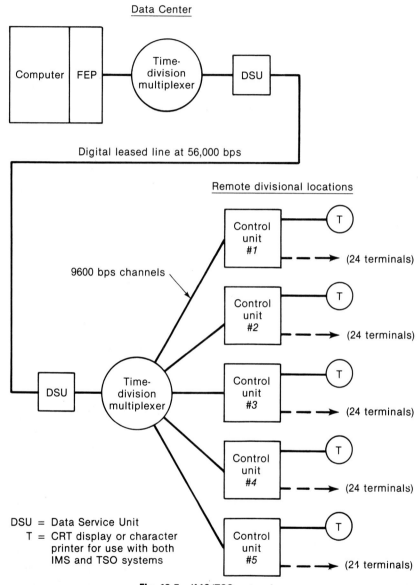

Fig. 13-7 IMS/TSO network

One printer is connected to each control unit, which can receive printed output in two different ways. A host computer application program can direct printing to any of the five character printers on the network in what is referred to as *on-line printing*. This is a purchase order, a report, or anything resulting from computer processing of data. One printer is connected to each control unit to provide each of the CRT display operators with the capability of printing any screen received on the display when a printed copy is desired. This *copy function* is restricted to the printer or printers connected to the same control unit as the CRT display being used. There can be multiple printers on the same control unit, which can be "customized" (programmed) to copy from certain CRT displays to certain printers. In the IMS system any CRT display or printer can send a short message to any other terminal on the system. The printers have or do not have keyboards, respectively keyboard-send-receive (KSR) and receive only (RO).

In the example each of the control units is limited to 32 terminals, but control units for only 16 or even 8 terminals are available. There is also a CRT display (IBM 3276) that has a built-in control unit and permits a maximum of eight total terminals to be attached.

In IBM's Network Control Program (NCP) for the front end processor, IBM 3705 or 3725, all applications on a host computer are made available to all terminals on the 3270 network. Thus an operator can log onto IMS for a while and then log off and log onto TSO. Some of the terminals can represent themselves as multiple logical units and remain logged onto multiple applications at the same time. Also available is a software package (MSNF) that permits a terminal to access applications on multiple host computers. At least one of the other front end processor vendors provides the multiple application access capability without requiring IBM's full NCP program, and there are other 3270-type control units that permit access to multiple computers. If a line is set up on the IBM 3705 or 3725 to run in the *emulation mode* (non-SNA/SDLC), multiple applications cannot be accessed and a host computer channel address is needed for each communications line and not just one address for the entire IBM 3705 as is possible under NCP.

A 56,000-bps digital line was selected for this network because it was less expensive than the comparable five analog 9600-bps lines and also more reliable and more easily maintained. The multiplexer was required to split the 56,000-bps physical line into the five 9600-bps functional channels. A statistical multiplexer was not selected for this application because the traffic volume is so evenly divided among the five channels. If some of the control units were used very lightly, we could get more throughput out of the single 56,000-bps line by installing statistical multiplexers at an additional cost.

14

Data Network Maintenance and Troubleshooting

Maintenance Resources

As with an automobile or any piece of functional equipment, the steps that are taken toward preventive maintenance and diagnostic maintenance problem resolution can predetermine the overall volume of network problems and the time and effort required to resolve the ones that do occur. Prior to even starting up a network, it is vital to plan and implement a network maintenance system, however simple it can be and however small the network can be. Items that should be considered for a proper network maintenance master plan are discussed in this chapter.

Telecommunications Personnel

Select and identify specific personnel to handle data communications problems. The use of computer operators at the data center to resolve communications problems can be workable if the network facilities are not extensive or complicated. The best approach, if affordable, is to designate one or more persons, full time if possible, who can be trained as telecommunications analysts and perhaps get involved with voice communications also.

Documented Maintenance Procedures

Devise a set procedure for attacking communications problems. Yes, many problems will be different from the previous ones and some will be very unique. The general procedure for investigating problems, however, can be predetermined and should be well documented as instructions for working on any communications problem. Prewritten instructions become particularly useful when the person who normally handles the problem is not available. Along with this, it is a good idea to also document any problem that is worked on in order to gather statistics for maintenance planning and, more important, to identify the symptoms in case the same problem should arise in the future (as it almost always does).

Line Monitor

One piece of diagnostic equipment that is practically a "must" for any in-depth network maintenance is a line monitor (see Fig. 14-1). It is a box with a small CRT display that shows the data characters as they come into or leave the terminal equipment. Some line monitors also have the capability of performing a bit error rate test (BERT), which is a very worthwhile option.

The monitor connects the modem and the terminal via an EIA (RS-232C type) cable extension. The bits are assembled into characters and displayed in either ASCII or EBCDIC code. Sometimes the received characters are shown on one line and the transmitted characters

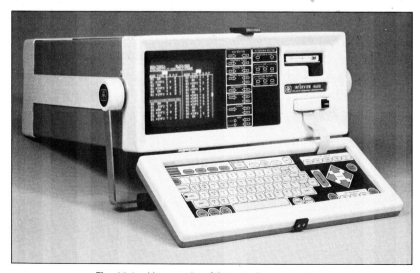

Fig. 14-1 Line monitor (*Atlantic Research Corp.*)

on the next line or a light background is used to differentiate between the two. The monitor has some buffer memory so any captured data can be scanned backward or forward for analysis. Magnetic tape cassettes and diskettes are available for increased memory storage.

The more expensive line monitors can be programmed to transmit certain protocol characters to test the response received against what is actually expected (DTE/DCE emulation). A knowledge of polling and protocol become very important in determining the source of a malfunction in the line control procedure. Some of the cluster controllers for CRT displays provide an extra maintenance board that permits one of the displays to show the traffic characters for that particular line only, similarly to a line monitor.

Response Time and Performance Analyzer

There are a number of inexpensive response time monitors available today that measure the total response time from when the data is sent to when returned data is received. Most of these devices connect to a CRT display with an optical coupler that merely detects when the INPUT INHIBIT light goes on (data outgoing) or goes off. Some of the devices also connect via the RS-232C cable by means of a special bridge connector. Some of these units have printers to accumulate statistics over an extended period of time (see Fig. 14-2).

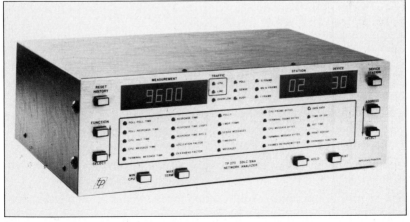

Fig. 14-2 Response time and performance analyzer (*Teleprocessing Products, Inc.*)

Most complaints by system users of excessive response time turn out to be due to delay in computer processing of the transaction and not to excessive data transmission time. There are several components that make up the total response time from when a user strikes the ENTER key on a CRT display to when returned information is received on the user's CRT screen. The cluster controller (IBM 3274, etc.) at the remote location can increase the response time if it is attempting to handle too many individual terminals, like 32. The front end processor (IBM 3705, etc.) can have a problem with its total maximum throughput. The modem and communications line can also be the cause of excessive response time with something like excess retransmissions generated by subgrade performance of the line or the modem. Remember that the communications line portion of the response time is usually about a half second for a half screen (1000 characters) of data at 2400 bps going to the host computer and a 50-character response being returned to the CRT screen. On the other hand, the host computer can require several seconds to actually process the information, including disk access time. Of course, it is important to verify which element of the group of network components is causing the excessive response time. System monitoring programs (software packages) can be run in the computer, or a hardware performance analyzer can give this breakdown of response time components to direct attention to the source of the problem.

There are more sophisticated performance analyzers that not only measure response times but also print reports concerning line utilization information. Statistics can be accumulated on line loading, effective baud rates, number of polls, character counts, and transmission errors—information which can be used to fine-tune or balance a communications network. These devices are tapped into the regular RS232C cable by using a Y connector.

Patch Panel

If multiple communications lines are involved, a patch panel should be considered. The digital (25-pin) patch panel permits the manual switching of modems between computer ports; the analog (two- or four-wire) patch panel permits the manual switching of modems between lines. Each provides a means of gathering all communications lines into one central box for a neater installation. It is a means of switching lines easily, if a spare line is involved, or testing with a similar line known to be good. It also provides a way to patch in a line monitor without interrupting any of the lines. Some patch panels have built-in line monitors, which usually cannot be transported to remote locations as a regular, portable line monitor can be (see Fig. 14-3).

There are much more sophisticated switching devices that can switch in backup front end processors or other host computers, provide port sharing, and offer many other capabilities.

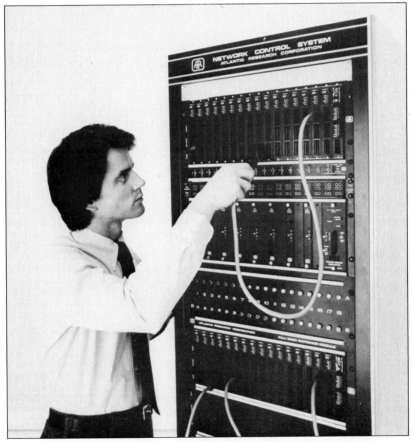

Fig. 14-3 Digital patch panel (*Atlantic Research Corp.*)

Network Control System

If the data communications facilities are quite extensive and critical to the operation of the business, consideration should be given to a network control system, including both hardware and software. Several vendors offer complete network control systems that have a network control computer processor between the host computer's front end processor and all the outgoing communications lines. Such a system would probably have an associated CRT display and a small page printer. Problems can be diagnosed and quickly corrected at both the central site and all locations. Many offer special modems that can be continuously interrogated by the central site network control processor and even uncover possible problem situations before they become seri-

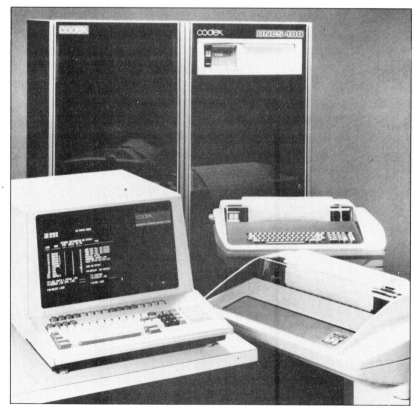

Fig. 14-4 Network control system (*Codex Corp.*).

ous. The AT&T Information Systems offering called Data Phone II is one of the modem-type network control systems. The modem contains microprocessors that can provide real-time diagnostic information to the central control modem at the network control site. A printer can be added to print out the diagnostic details, and a CRT display to request specific information and tests of the system also can be added.

The more sophisticated network control systems provide statistical reports for line-loading analysis and network performance studies. Lines can be switched to bring in backup lines. Network control consoles are available not only at the control site but can be made available at other network nodes also. Series of tests can be performed to check out suspected error conditions (see Fig. 14-4).

There are also software packages like IBM's NCCF, NPDA, CNM, SSP, and NDLM that run in the host computer and provide some of the network control functions such as diagnostic interrogation of the modems, line testing, and gathering of statistical information.

Voice-Grade Line Problems

Lines were installed for voice conversations long before computer data transmissions appeared on the scene, and it has seemed logical that the same lines could be used for the transmission of data. The actual line facility for a voice call can be a combination of twisted pairs of wires, microwave radio links, coaxial cables, fiber optics cables, and satellite circuits. The final connection to the ultimate user at each end of the chain is normally a twisted cable pair connecting the customer's station equipment to the telephone company's nearest central office. The normal two-wire switched telephone channel has a bandwidth of 300 to 3400 Hz (cps), though up to 12 channels could be available with one physical twisted pair line. A switched voice channel can normally support data transmissions up to speeds of 1200 bps, and even higher speeds are possible by using special technology.

A four-wire dedicated or leased channel is called a No. 3002 channel in this country. It can normally support data transmission speeds of up to 9600 bps without any special line conditioning or special modem features and up to 14,400 bps and higher by adding amplitude and delay equalization conditioning. We will look at some of the factors that affect the data transmission capabilities of a voice-grade line and increase the error rates if not dealt with properly.

SIGNAL POWER LOSS

Attenuation refers to the loss in power as a signal travels over various elements of a communications line. This loss of magnitude or power level is usually expressed in decibels. You can calculate decibel losses or gains by using the following equation and a pocket calculator.

$$dB = 10 \log \frac{P_{out}}{P_{in}}$$

where dB = decibels
  P_{out} = output power, in watts
  P_{int} = input power, in watts
  log = logarithm of (P_{out}/P_{in}) to base 10

Some pertinent facts about decibels are:

(1) A 1 · dB change in signal level is the least discernible change that the human ear can detect.

(2) A negative number of decibels indicates a loss in signal level. For example, if the power level drops by one-half, the result is a loss of approximately 3 dB. If the power level drops to one-tenth of the initial value, there is a 10-dB drop in signal level.

(3) A positive number of decibels indicates a gain in signal level. For example, each time the signal level is doubled, the resulting gain is approximately 3 dB. Each signal increase by a factor of 10 represents a gain of 10 dB.

The telephone industry uses decibels referenced to 1 milliwatt (mW) of power for most signal level measurements. Thus, 0 dBm is defined as 1 mW of power. Table 14-1 shows typical values of power levels referenced to 0 dBm.

The maximum permissible loss as a signal travels through the circuit from origin to final destination has been specified as 16 dB. For data transmission at 1200 bps and at a frequency of 1000 Hz the attenuation, or loss in decibels, is normally about 10 dB. Each element of a circuit generates its own loss number, and the loss numbers are cumulative and add up to a total loss for the circuit. The attenuation will differ with frequency on the same communications channel. Amplitude

Table 14-1 Power Levels Referenced to dBm.

Power Level (watts)	dBm	Power Level (watts)	dBm
1	30	0.0005	−3*
0.1	20	0.0001	−10
0.01	10	0.00001	−20
0.002	3*	0.000001	−30
0.001	0		

* Approximate value.

distortion of a signal is caused by a variation in attenuation over a range of frequencies. For instance, when a modem that uses frequencies from 800 to 2500 Hz is employed, measurements should be made at each 100-Hz frequency interval to make sure that the attenuation is acceptable at all frequencies. Also, the measurements must be made on data being transmitted in both directions, because attenuation can be acceptable from station A to station B but not in the reverse direction. Line conditioning performed by the telephone company can help resolve an excessive attenuation problem. Type C conditioning controls attenuation and envelope delay distortion. Type D conditioning controls the signal-to-noise ratio and nonlinear distortion.

ECHOES
Echoes are another problem encountered in the transmission of data over voice-grade lines. They are caused by an unmatched impedance in a communications circuit. The original signal is reflected with enough magnitude and delay to cause some confusion to the users of the circuit. A way to resolve the problem in one-way transmission is to install an echo suppressor, which prevents any signal from being sent back the other way by inserting a 50-dB loss in the return path. If full duplex operation is desired, such an echo suppressor must be disabled. Another way, particularly in a satellite circuit, is to use an echo canceler, which develops a signal comparable to the echo to be subtracted from it and thus cancel it.

RETURN LOSS
Return loss at the point of mismatched impedance in a circuit is another possible source of problems. Any return loss below, say, 42 dB

between 800 and 2300 Hz could cause a problem. A high return loss indicates a good impedance match, and a low return loss indicates a poor impedance match which should be corrected.

NOISE

Noise is caused by undesirable misdirected signals. The term *white noise* refers to a hissing which appears on a circuit when no traffic is being carried. The noise is expressed in decibels as a signal-to-noise ratio. This type of noise, as compared to another type called *impulse noise,* is rather constant in magnitude. The level of noise on a communications channel can be measured by comparison to a signal level in a test channel and expressed in decibels as a certain signal-to-noise ratio. The noise generated by each link of a multilink communications line is cumulative in an analog transmission. Thus the greater the number of individual links, the more noise is generated and the smaller the signal-to-noise ratio. The result is a poorer quality of transmission. This cumulative effect is not applicable to digital transmission, in which repeaters are utilized to regenerate only the original signal and not the noise. Signal-to-noise ratios in analog circuits can be improved by the use of compandors. These devices compress the dynamic range of voice signals received and later expand them to their original range; the result is a reduction in noise level. A signal-to-noise level of 24 dB or greater is normally acceptable.

Impulse noise is characterized by its nonconstant magnitude and random high peaks of short duration. It can be a critical problem with data communications transmissions, but it does not normally interfere with voice transmissions. The telephone company has defined limits for the amplitudes of impulse noise spikes on the line; they are from about -6 dB to $+2$ dB with reference to the received signal. Measurements are normally concerned with the number of acceptable noise pulses within a given period of time. For example, no more than five noise impulses are allowed on a line if the impulse noise amplitude is about 2 dB above the received signal amplitude.

CROSS TALK

Cross talk is the situation in which one channel picks up a signal traveling in a closely associated channel. Most of us have heard cross talk during voice conversations, but it can occur in data transmissions also. The problem can be resolved by the correct assignment of control and signal circuits.

DELAY DISTORTION

Delay distortion is caused by filters and loading coils in voice channels that create phase differences between different transmission frequencies. As transmission speeds increase, bit lengths decrease and the data at different frequencies require varying periods of time to reach their destination. The result is called delay distortion. Equalizers are used to compensate for delay distortion in a channel. In a digital line the regenerator reduces the equalization problem and only a simple digital filter can be required to combat delay distortion.

CARRIER SHIFT

A change in frequency tone on a voice-grade line, called carrier shift, is caused when the original signal frequency is translated to a higher frequency for transmission and then lowered to a voice or data frequency at destination. A tone originally transmitted at 1120 Hz could actually be received at 1130 Hz. The allowable carrier shift is ± 7 Hz, and anything more than that could result in unrecognizable voice or data information.

PHASE JITTER

Phase jitter concerns a variation in the time of tone transitions from the point at which the signals are received. This occurrence is not serious for voice transmissions, but it is for data transmissions, particularly at high speeds. When the phase shift from the point of origin to the destination exceeds 15°, data transmissions can be impaired.

HARMONIC DISTORTION

Harmonic distortion is related to the incorrect reproduction of a pulse signal as it is transmitted over a communications circuit. For example, a square-wave pulse signal requires a bandwidth much greater than the frequency at which the square wave is transmitted. The signal contains sinusoidal (sine-wave) components of varying amplitudes at frequencies of odd multiples (harmonics) of the fundamental frequency. (Other nonsinusoidal waveforms, such as triangular pulses, will possess both even- and odd-harmonic components.) In order to transmit the original signal waveform, the communications channel bandwidth must be at least wide enough to pass a sufficient number of harmonics to recreate the signal at the receiving terminal. For example, if a square-wave signal with a frequency of 1000 Hz is to be transmitted, a bandwidth of about 5000 Hz to accommodate the third and fifth harmonic signals will be needed for a reasonable reproduction of the square wave at the receiving terminal. Thus we see that a narrow bandwidth could cut out the early harmonics and result in a distorted signal.

Nonlinear Distortion

Nonlinear distortion can be caused by amplifiers and other devices generating undesirable signal components that can result in phase distortion, etc., which can not harm voice transmissions but can cause serious problems for data transmissions at speeds of 2400 bps and greater.

A change in the shape of a signal when it is received at its destination is called bias distortion. Characteristic distortion is caused by a new signal waveform arriving at the destination before the preceding wave has been properly registered at the receiver.

General Problem-Resolving Procedure

Many data communications problems fall in a general category, such as open line or defective modem, and will be identified by following the steps of a general procedure for problem solving. Many others will turn out to be quite involved and can even be "firsts" with no previous documentation. In either case, the maintenance procedure steps stated below should be of value. In the final analysis, however, it is the ingenuity of the maintenance personnel that gets the network back up and running.

Verify and Identify the Problem

The first step in resolving a data network problem is to verify the complaint and make sure the problem actually exists. An inexperienced operator at a remote location can sound the alarm if the brightness control on a CRT display has been turned down or something is unplugged. Check the status of other terminals at the same location or on the same line to determine whether it is a terminal problem, line problem, or network problem. One good rule to keep in mind when confronting data communications problems is that you should get the facts first hand and not take anybody's word for anything. This can hurt someone's feelings, but when you have been burned a few times with hours of unnecessary work, the skepticism proves to be worthwhile.

Isolate the Problem

Data communications networks are subject to a variety of problems that could interfere with their functioning properly. Since so many interrelated factors are involved, it is extremely difficult at times to pinpoint the true source of the problem. One approach that seems to be very popular with many information systems managers is to call in all of the vendors to resolve any data communications problems that are not readily identifiable. Getting all of the vendors in at the same time is not only difficult to schedule but is usually an unreasonable waste of manpower. Many vendors have become so perturbed by this approach that, in self-defense, they have established policies whereby the customer is charged for a normally free service call if the problem turns out not to be theirs.

We should be aware that the primary objective of an inferior maintenance person is to prove that the problem is not the fault of his or her company, and there are usually lots of indicators available to fortify this point of view. A superior maintenance person attempts to find the source of the problem no matter where it may be situated. The point stressed here is that it makes sense to place the responsibility to isolate the problem (to a particular vendor) with the user or customer.

Contact the Computer Data Center

Data communications lines are usually controlled by computer operators at the computer data center, or there can even be a separate network control facility available. Either way, it would be advantageous to inquire as to the status of the line or terminal in question.

Sometimes a terminal, cluster controller, or line that is in trouble can be "varied off line and then on again" to resolve a possible software problem. If the entire system is in trouble, a shut down and restart can be in order. A front end processor can also be used to reset a line that has a problem and can supply diagnostic information concerning the line.

Another source of information is the data event control block for the communications line software such as IBM's BTAM. The error code for a line abort can be taken off the computer listing and checked against the associated error listing for a description of what the computer thinks went wrong.

In any case, the computer operators at the data center should be made aware of the situation so they can notify other users or check other lines, and gather useful diagnostic information.

Check the Modems

If all of the terminals on a line are inoperable, it might be a good idea to look at the modems (or DSU) or any multiplexers that may be involved. If the CARRIER DETECT light is out, you have a line or modem problem. A "streaming modem," which is hung up during the sending of request to send (RTS) signals, can be disastrous on a polled multipoint network. There is available a black box that will recognize and shut down a streaming modem automatically, but a look at the modem's RTS lamp should at least pinpoint the problem.

Most modems provide both a self-test and a loopback test of the entire line. Keep in mind throughout this discussion that not even a device that passes a test can be ruled out completely as the possible source of the problem. Vendor equipment tests are not all-inclusive, and they may not even be testing the problem that actually exists.

Check the Line and Contact the Telephone Company

There are small audible signal speakers similar to those of a telephone handset that permit a person to hear the tone of the data to make sure it is coming through. A breakout box, shown in Fig. 14-5, could be inserted to test the EIA pin voltages to make sure the terminal and

Fig. 14-5   Line breakout box tester (*Black Box Corp.*)

modem are sending the correct signals. Some of these boxes also perform a bit error rate test (BERT), which indicates whether the line is
generating excessive errors. Much more sophisticated equipment is
available for checking communications lines, but its use can be justified
only in exceptionally large installations and by trained line maintenance personnel.

Most companies find it best to leave any detailed testing of the line
to the telephone company or common carrier. They can run loopback
tests from the maintenance center. This is particularly effective with
digital lines where the tester can check a cross-country circuit while the

customer stays on the original complaint telephone. If the problem persists, and it very often does, the next step is to request an end-to-end test by the phone company. The company will dispatch a repairman to each location where more extensive tests, such as line decibel loss, can be performed. If the tester still can't find the problem and all indications point to a line problem, the phone company's local Datec division can be contacted. These are the data maintenance experts for the Bell System.

Use a Line Monitor

When there is a problem with a communications line, it would be nice to "just get in there and see what's going on." The line monitor performs this function to a good degree, but, like any piece of test equipment, it does not detect all of the problems that could occur. It is very good for things like the use of an incorrect polling address, which could have been changed at the host computer or the remote terminal since the last successful transmission.

If either the host or remote terminal fails to respond to a command sent by the other, that device should be questioned, but don't assume that it is absolutely the unresponsive device's fault. The other location could have failed to send the proper RS-232C clear-to-send signal. Certainly incorrect and garbled characters seen on the line monitor are a good indication to look for the cause of the problem. Any failure to follow the expected line protocol should be investigated.

Cluster controllers often send out status messages when they have a problem such as a busy condition or a detected error in the device itself. The status message reads something like "SOH % R STX," but the actual hexadecimal characters seen on the line monitor must be coded by using the conversion table shown in Table 3-1 or 3-3.

The "What Has Been Changed" Approach

Data communications line problems can become extremely difficult to solve because of the wide variety of sources of error such as software (application program, communications monitor, access method, operating system, etc.), hardware (computer, front end processor, modems, lines, terminals, etc.), and people (computer operators, programmers, systems analysts, terminal operators, communications analysts, etc.). A telephone company in Hartford, Connecticut can decide to use satellite lines for its dial-up network during only part of the day, which could prevent getting a connection for a terminal that has otherwise been operating satisfactorily for the past week.

When the resolution of a data communications problem becomes frustrating, ask yourself (and everyone else) "What has been changed?" You might find out that the computer room is using an outdated version of a communications program or some other such unlikely event.

"Ye Old Swapout" Method

There are a lot of managers in data processing who consider the practice of swapping a suspected piece of equipment with a known good piece of equipment as being unscientific and too uncertain. Certainly a logical diagnostic approach would be preferred, but it can take too much time and might not result in a timely solution to the problem. We would venture to say that more data communications problems have been solved in a shorter period of time by swapping out equipment than by any other method.

Keeping spare equipment on hand can be economically justified, particularly if large numbers of the same item are in use. If the equipment is the source of the problem, swapping it is the only way to get back up and running immediately, since most maintenance people require a half day or so to appear on the scene. Swapping can also be done if there are no spares but two communications lines have the same elements. One user might agree to a shutdown of 5 or 10 min to help find a problem with another user's equipment. This works well with modems, CRT displays, telephone company line terminators, multiplexer channel boards, communications leased lines, dial-up ports, and many others. The method is particularly convenient for checking the line adapter or port on a front end processor, which often can be swapped via a command on the device's console.

About the only way to be certain of the cause of a data communications problem is to be able to recreate it at will. This is often a good step to take, particularly when swapping out equipment, to make sure the actual cause was not a faulty connection or just disappeared by itself.

Call in the Vendor

Of course, you may have to eventually call in a vendor to check out equipment, but this step has been left as the last choice because of our firm belief that the problem must be isolated to a specific vendor if at all possible. If all of the steps of a problem-solving procedure are properly taken, chances are that the vendor called in will not spend time attempting to place the blame on someone else.

15

Voice Network Facilities

The voice networks include all of the telephone instruments, switch-boards, and telephone lines that we use each day to talk to people who are either in the next office or miles away. Voice networks were, of course, around many years prior to the use of the voice telephone network for the transmission of data. Telegraph and TWX and TELEX also go a long way back, but they were used for sending messages and not computerized data as they are today. Since the same lines that are used for voice transmissions are often used for data and mixture of the two for transmission together is becoming more prominent, we should be aware of the structure of our voice network.

The operation of today's public dial-up or switched network is controlled by signaling, which consists of electrical pulses that are used to manage the network. It is employed to connect users to the network, indicate a busy station, ring an open station, provide dial tone to indicate readiness to receive, initiate the billing system and, finally, disconnect the user. Another type of signaling is needed to control the circuits between the telephone company's branch switching offices as the call progresses beyond the user's local central office. Signaling can take the form of pulses for rotary or touch-tone dialing, audible tones to indicate a dial tone (availability) or busy condition, a ring to alert the distant location, or a signal to indicate whether the location is in the on-hook or off-hook status.

The local telephone company central office becomes aware that a user wants to dial a call because of a change in the circuit voltage when the handset is lifted. The central office then provides a dial tone as an indication that the circuit is available for use. The user's dialing creates 48-V pulses that activate the switching equipment at the central office. With touch-tone facilities, the depression of a push button causes the emission of two of a possible eight audio-frequency tones. Each digit has its own two-frequency tone combination. The central office equipment interprets the tones and stores the digits in a register for further processing of the call to its destination. The ringing at the destination is caused by a 100-V or so signal at 20 Hz. Though there are many other common carriers today, we should be familiar with the Bell System Network, since it is the basis for this country's public voice and data communications network.

Bell System Network

The local telephone facilities throughout the country have been initially divided into 161 separate geographical areas known as local access and transport areas (LATAs). Each of the 22 Bell Operating Companies (known as "the telephone company") have been given control of the intraLATA (within the area) traffic within that BOC's boundaries on the old tariffed and federally regulated basis. The interLATA traffic is to be provided by the newer unregulated and competing interchange carriers such as MCI, GTE Sprint, AT&T Communications, etc. Each of the Bell Operating Companies has its own territory, or portion of the United States. The physical lines themselves can be made up of any combination of microwave, satellite, or regular two- or four-wire lines, and the composition can vary with the time of day.

THE DIAL-UP NETWORK

The dial-up network's initial point of interfacing with the homes and business establishments is the local central office (CO). The four-wire line, or bundle for businesses, connecting the customer to the central office is called the *local loop*. Each central office is connected to several *toll offices* via lines called "trunks," which act as switching centers as the path of the dial-up call progresses across the country to its final destination after going through the customer's central office and local loop. Since the toll offices are numerous, there could easily be 25 different routes that cross-country calls could ultimately be switched into, depending on the individual line loading out of each of the toll offices. That is why, when direct distance dialing (DDD) is used, it is recommended that another call be placed if the one originally placed is unacceptable because of excessive noise or for some other reason. The progress of Bell's switching mechanisms from step switching, to crossbar, to the latest electronic switching systems (ESS computers) has been sporadic around the country, so we can expect to see all three switching methods in operation, depending on where we are calling. Frequency-division multiplexing is used on a phone line twisted pair of wires to provide 12 normal voice channels of about 3000 Hz (cps) each. In order to accomplish this more effectively, the frequency levels are raised to the 60,000- to 108,000-Hz range, but the frequency presented to the person at the other end must be reduced to the 300- to 3300-Hz range in order to be recognized by the human ear.

THE LEASED LINE

The leased, or private, line handles voice, data, television, etc., for customers whose requirements are great enough to warrant a dedicated link between two or more locations. Each leased line has its own local loop to the central office, and from there permanently installed dedicated lines connect between all of the toll offices along the route to the final destination, again via that location's central office with a local loop to the place of business. Since the same route is used each time a transmission is initiated, a leased line can be "conditioned" for data use to operate more effectively. Unlike the dial-up facility, the cost of which varies with the distance and the duration of the call, a leased line has a fixed monthly charge that varies only with the distance (and any extra conditioning), so loading the line day and night could be an economic advantage.

Available Voice Communications Services

Local Dialing

Local dialing constitutes the local calls that are made within a town, city, or metropolitan area. There is an option concerning the monthly billing for the service, which depends on the anticipated usage. If the phone is to be used primarily for incoming calls and few outgoing calls, a *measured service* would be advisable. If a considerable number of calls are to be placed, it would be best to arrange for a nonmeasured service that includes a number of units to be used for the bulk of the calls anticipated for the month. A unit amounts to $0.0529, and perhaps one to four units are required for most local calls, depending on the distance and duration of the call.

A *metroline service* would be advantageous when large numbers of short calls are to be made within the metropolitan area of a large city. This is a flat rate for unlimited calls within the designated area. There is also a *metropolitan service;* a flat rate is charged for all calls within the same local or central office area. In small communities and towns there would probably be a flat rate for all calls within the area plus charges by units for calls outside the area. In a large city a billing system whereby one unit is charged for any call within the city, no matter how long the call lasts, might be available.

Direct Distance Dialing

The operation for DDD is very similar to local dialing except that a 1 and the area code are needed and the billing method is quite different. The customer is charged by the minute (for whole and fractional minutes) for each call. Rates are reduced during other than regular business hours. Though this method of calling is more economical than collect or credit calls that require operator assistance, any volume of use should dictate a study of less-expensive alternatives. The use of one of the less-expensive long-distance companies such as SPRINT or MCI should be considered.

Wide-Area Telephone Service

The WATS service was conceived by the Bell System to give a lower bulk rate to businesses that require a high volume of phone calls. It is now called a *measured WATS system* in that the customer pays a reduced rate for a preselected number of hours of calling time per month. This method of customer billing affords a sizable price decrease as compared to regular direct distance dialing. Both intrastate (within the state) and interstate WATS services are available.

In the interstate WATS service the country is divided into six areas each of which includes several states (see Fig. 15-1). The cost of an hour's calling varies from the lowest charge for the closest area

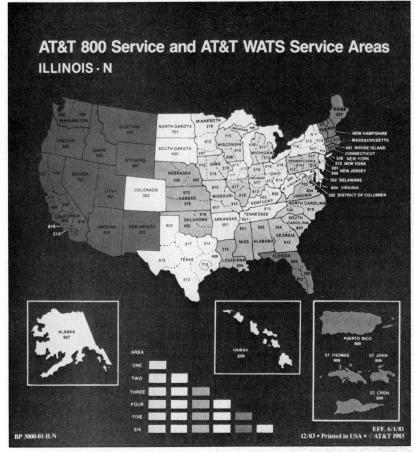

Fig. 15-1 WATS map (*AT&T Communications*)

(adjacent states) to the highest charge for the farthest area. The service can be reversed so customers in one or all of the six areas can call a central location via IN-WATS at no charge to the caller. This is often referred to as the 800 area code service.

It should be remembered that both WATS and IN-WATS represent only a method of billing for use of regular long-distance telephone facilities. If it is necessary that five conversations take place at one time, five individual WATS lines will be needed. Appendix D shows the pricing and other details of AT&T Communications' WATS service. WATS-type services are also offered by MCI (Microwave Communications, Inc.), ITT (International Telephone & Telegraph), and SPRINT (GTE Sprint Communications), possibly at a lower cost.

Foreign Exchange Line

The FX line is a dedicated voice line similar to a leased line for data except that it is "open at one end." Since the FX voice leased line connects to the switched network, one of the two ends can be dedicated as a regular telephone number in that local central office network. This means that, at the "open end," any telephone in that area can place and pay for only a local call to talk to the telephone station at the other end of the FX line, though it can be hundreds of miles away. Thus, as shown in Fig. 15-2, any telephone in Denver can place a local call to talk to the single telephone (or switchboard) in Chicago headquarters. Headquarters in Chicago can only talk to the single original

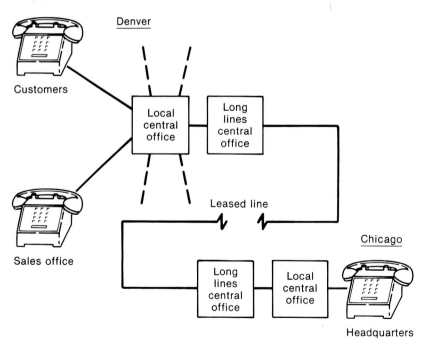

Fig. 15-2 Foreign exchange line

telephone (or switchboard) at the sales office in Denver at no additional charge, as with a dedicated data line. In effect, Chicago headquarters has a local phone extension in Denver.

Telpak Services

Telpak is another high-volume offering by the Bell System and Western Union. Telpak C consists of 60 voice channels at full duplex on a dedicated basis (leased lines), and Telpak D includes 240 voice channels. Telpak is currently available only for intrastate (within the state) applications. The channels can be used for both voice and data.

T-1 Carrier

The T-1 Carrier is used quite extensively by the Bell System in its voice network around the country. Its use as a customer-offered digital leased line facility is fairly recent. Through the use of pulse code modulation, an analog voice signal can be converted to a binary pulse stream similar to a digital data stream. The "digitized voice" stream is then regenerated at frequent intervals to strengthen the original signal without amplifying any distortion that was present. By using this technique, the capacity of a single twisted wire pair (T-1 Carrier) can be expanded to 24 pulse-code-modulated voice channels, or 1.544 Mbps of data. Of course, the binary pulses must be converted back to analog signals at the destination for voice communications. A T-2 Carrier can accommodate 96 voice-grade channels, or 6.312 bps, and a T-4 Carrier can transmit 4032 voice-grade channels, or 274,000,000 bps of data. At the time of writing AT&T had plans to double the capacity of the T-1 Carrier only for speeds of 4000 bps and below. That would add up to 20 more voice channels (or voice-grade data channels), making a total of 44 digitized voice channels and 4 signaling channels, all operating at 32,000 bps.

Time-division multiplexing is used to pass the frames of binary pulses at the full line speed and send alternate frames from various telephone conversations or computer data if a mixture is desired. Each channel sends a frame containing eight bits which travels at a speed of 8000 frames per second.

Time Assignment Speech Interpolation

Time assignment speech interpolation is a special technique employed by the common carriers to combine multiple voice conversations into a single channel, particularly for overseas calls for which the channels are expensive. Advantage is taken of the fact that there are lulls in telephone conversations and only about half of a full duplex voice channel is being used during a conversation. The actual channels are swapped between users so quickly as to be hardly noticeable, and thus more voice conversations can be accommodated by this method of more fully utilizing the complete capacity of the channels.

Voice Line Loading

The amount of communications traffic that can be loaded into a single communications channel or line must be given very serious consideration when a new network is planned or an existing network that has increased its traffic volume over the months is reevaluated. Though it is true that existing communications links can be expanded as traffic volume increases, it is far better to make a proper estimation of the requirements in advance and thereby avoid any overloading that could be detrimental to the company's business. About the only situation in which line loading is not a major concern is that in which we are using local or direct distance dialing and each phone station has its own channel to the phone company central office. Even with WATs, which is a dial-up system, we must analyze the traffic load to make sure the optimum number of lines are selected to give the greatest economic advantage. Dedicated voice lines, like FX lines and PBX tie lines, are often installed in multiples.

One important fact needed in the installation of dedicated (or WATS) lines is the volume of traffic anticipated. A good source of information concerning traffic volume consists of previous long-distance phone bills. Local or unit call information can be obtained by requesting a magnetic tape record from the phone company, which would have to be extracted by writing a computer program to create a report or contacting one of the small companies that offer that service. There are other sources, including PBX call detail recording or even a rough estimate of future activity expected. In any case, we are looking for the seconds of usage, by hour of the day if possible. Any final line configuration should be able to handle the peak traffic for the day, but maybe the users will have to wait a little longer to get a call through.

Dialing and getting an occasional busy signal would seem acceptable in most situations, since this approach assists in avoiding excess, wasted line capacity. On the other hand, successive busy signals and

Table 15-1   Line Capacities.

Lines	Grade of Service									
	P01	**P02**	**P03**	**P04**	**P05**	**P06**	**P07**	**P08**	**P09**	**P10**
	CCS per hour									
1	.4	.7	1.1	1.5	1.9	2.2	2.6	3.0	3.4	3.8
2	5.4	7.9	9.7	11.3	12.9	14.2	15.5	16.8	18.0	19.1
3	15.7	20.4	24.0	26.9	29.4	31.7	33.9	35.9	37.8	39.6
4	29.6	36.7	41.6	45.7	49.1	52	55	58	60	63
5	46.1	55.8	61.6	66.6	70.9	75	78	81	85	88
6	64.4	76.0	82.8	89.3	94.1	99	103	107	110	113
7	83.9	96.8	105	112	118	123	128	132	136	140
8	105	119	129	137	143	149	154	159	163	168
9	126	142	153	162	169	175	181	186	191	195
10	149	166	178	188	195	202	208	214	219	224
11	172	191	204	214	222	230	236	242	248	253
12	195	216	230	240	249	258	264	270	277	282
13	220	241	256	267	277	286	292	299	306	311
14	244	267	283	295	305	314	321	328	335	341
15	269	293	310	322	333	342	350	357	364	370
16	294	320	337	350	362	371	379	387	394	401
17	320	347	365	378	390	400	409	416	424	431
18	346	374	392	407	419	429	438	446	455	462
19	373	401	420	436	448	458	468	476	485	492
20	399	429	449	465	477	488	498	506	516	523
21	426	458	478	494	507	517	528	537	546	554
22	453	486	507	523	536	547	558	567	577	585
23	480	514	536	552	566	577	589	598	607	616
24	507	542	564	582	596	608	619	629	638	647
25	535	571	593	611	626	638	650	660	669	678
26	562	599	623	641	656	669	680	691	700	710
27	590	627	652	671	686	699	711	722	731	741
28	618	656	682	701	717	730	742	753	763	773
29	647	685	711	731	747	761	773	784	794	805
30	675	715	741	762	778	792	804	815	826	836

the resulting user frustration must be avoided. To obtain the correct balance between "wasted" and "overloaded" voice lines a grade of service analysis should be made. To make it, the actual expected minutes of use of a line for each hour of the day, or the average hour if the peaking is not severe, should be converted to CCSs (100 call seconds) by multiplying by 0.6. The resulting figure is the CCS per hour shown in Table 15-1. To use the table, decide the highest (poorest) grade of service that is acceptable (e.g., P07 means that 7 percent of the users will be denied a connection on the first attempt to dial out). Then search in the column of CCS per hour numbers to find the number closest to yours. Look over to the left column to obtain the number of lines required for your particular load (CCS per hour) and grade of service. As an example, if the lines must handle 120 min of activity each hour (120 × 0.6 = 72 CCS) and P02 grade of service is desired, then six lines are needed. Line capacity tables can also be based on Erlangs (i.e., a fully loaded line, or 3600 call seconds). It should be noted that increasing the number of lines from one to two at P01 permits 13 times as many call seconds.

Private Branch Exchange

The term PBX was derived from the original manual switchboard, with which we are not concerned in this discussion. We are talking about the "automatic" switches that do not require operators with patch cords but may have one or more operator control consoles (cf. Figs. 15-3 and 15-4). The new automatic PBX could be defined as a computerized switch that is used to interconnect all the telephones of a company's office, building, or complex to each other as well as to the outside world. The company's own telephones are normally connected to the PBX with two or three twisted wire pair phone lines, and the PBX, in turn, is connected to the nearest phone company central office with twisted pairs of wire phone lines, usually in larger cable bundles. The PBX processor contains memory and instructions that are either "burned into" computer chips or, in some cases, loaded in as a program similarly to a business computer operation. Some of these switches handle only regular analog voice signals, but most of the newer ones are digital devices that can handle both analog and digital signals. The PBX switch itself can be installed off-site at the phone

Fig. 15-3   PBX control unit (*GTE Business Communications Systems, Inc.*)

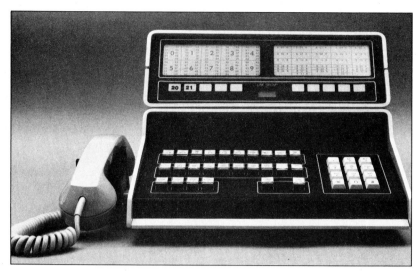

Fig. 15-4   PBX attendant console (*GTE Business Communications Systems, Inc.*)

company's central office as was done with most of the present Bell Centrex systems. However, the current practice is to install the switch on the company's premises, which seems to afford several advantages. Figure 15-5 shows examples of a Centrex off-site switch and an on-site switch. Note that the off-site switch requires a voice channel for each of the 50 phones at the place of business, whereas the on-site switch can operate well with a ratio of something like six phones for each channel needed to the central office. This advantage would disappear if the direct inward dialing feature were needed; then each phone would have its own voice channel anyway. PBXs, often referred to as CPE (customer premise equipment), are usually rated in size by the number of individual users, or stations, that they can support as well as the percentage of blocked calls, if any, that the switch cannot handle under a full load. Some significant PBX features are described below.

Attendant Display Console

The attendant display console, with its push buttons and display lights, is used by a company operator to control the PBX system and handle calls for employees who are not at their desks. It shows the calling source, called number, class of service, status of called exten-

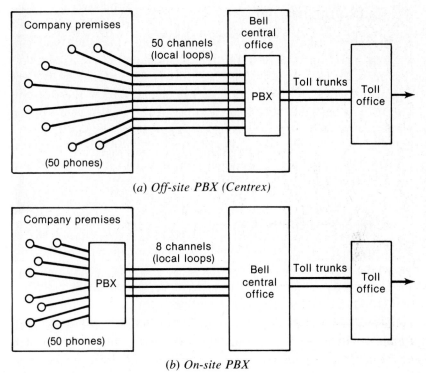

(a) *Off-site PBX (Centrex)*

(b) *On-site PBX*

Fig. 15-5 Comparison of on-site and off-site PBX

sion, automatic re-calls, etc. Normally, one or two of these consoles are required, but it is possible to operate without one in certain situations.

Direct Inward Dialing

The direct inward dialing feature provides each telephone station with its own phone number so that outside business associates can dial the phone station directly without intervention by the company's telephone console operator. The phone company provides a detailed listing of long-distance call costs by station. There are additional costs for this feature, particularly that of the increase in local loop channels to the central office because each phone station requires its own channel. Direct inward dialing can be worthwhile if the company employees are heavy users and are at their desks most of the time to take the incoming calls.

Route Optimization

Route optimization is also referred to as automatic route selection because, when an employee dials an outside number, the PBX makes the decision as to what route or outside line is to be used (e.g., WATS, FX line, or DDD). If there are a variety of outside lines to choose from, this feature becomes very important. It makes the decision for the employees, makes sure the correct line is actually used, and eliminates the need for two-digit access dialing. (This can also be accomplished using a WATS BOX or by running dedicated lines to a long-distance vendor, like MCI.)

Call Detail Recording

Call detail recording provides detailed information concerning the use of the system, such as a listing of all long-distance calls by individual phone stations. It could be very useful for budgeting, charge-backs, and monitoring for misuse of the system. It is a little expensive and perhaps should not be installed if the information is not important or if the information needed can be obtained by some other means.

Call-Back Queuing

If a station calls for an outside line (WATS, FX, etc.), and none is available, the station will be rung back when a line does become available. Call-back queuing is offered at no charge with some PBXs, and it is very helpful in fully loading the more economical outgoing lines.

Speed Dialing

The speed dialing feature permits quick dialing with two to four digits for commonly dialed numbers. It is available on most PBXs at no charge.

Electronic Phones

Electronic phones are special multibutton phones that offer a variety of additional features such as a hold button, access to FX lines, multiple incoming lines, and an intercom system. They cost more than the regular phone instruments and are usually limited to key personnel.

Multibutton Phones

Multibutton phones, often referred to as key sets, have up to six push buttons that permit access to up to five outside lines, plus the option to put a call on hold. There is an additional charge for key phones, and they have recently been superseded in most applications by the newer and more flexible electronic phones.

Toll Restrictions

The toll restriction feature can be used to limit certain outside calls by station, or totally (e.g., WEATHER), and it can be implemented by toll line, exchange, or specific phone number.

Call Transfer

A call can be transferred to another phone station without operator assistance.

Consultation or Three-Way Call

A party can be put on hold while a call is placed and completed with a third party, or the third party, or even more parties, can be brought into the conversation.

Group Pickup

Any station in a group can answer a ringing phone.

Night Service

Provides a universal chime for all phones at night, or provides certain night lines that are dialed by code, or provides dialing to a specific station only.

Call Waiting

An incoming call waits at a busy station, which later rings when the handset is hung up. (Some companies have installed a telephone answering device at each employee's desk to record any short incoming messages.)

Attendant Camp-on

An incoming call camps on a busy phone station, gives a short notification tone, and then connects when it can. If not answered within X (optional) minutes, the call goes back to the console operator.

Class of Service

Phone stations can be assigned to up to 16 classes for the purpose of restricting the use of outside lines, etc.

Three-Pair Universal Wiring

Three-pair wiring provides the new universal arrangement of pairs of twisted wires for each phone station except key sets, which require a larger cable.

Remote Access to PBX Services

Outside parties can dial into the PBX and use the special outgoing lines such as WATS and FX.

Direct Group Calling

Incoming calls can be sent directly to a group of phones from multiple outside lines. The calls hunt past any busy station.

Call Forwarding

A phone station user can arrange for all incoming calls to be directed to another phone.

Call Hold

A station user can put a call on hold by merely pushing a special button or the cradle button.

Centralized PBX Network

Under certain circumstances it could be economically advantageous to combine several separate company locations into a single PBX communications network with a traffic directing center. Figure 15-6 shows

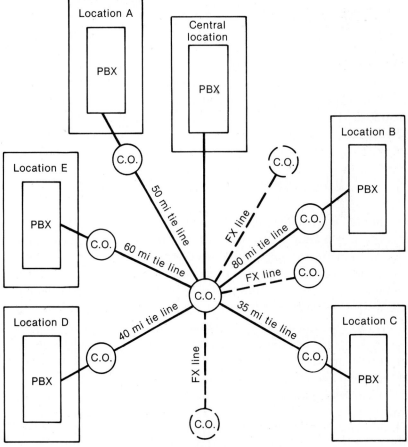

Fig. 15-6 Centralized PBX network

an example. Tie lines are used to connect to phone company central offices; they are normally installed in multiples so more than one call can be handled at a time. It should be noted that an FX line must be installed at a central office that passes on a lot of the network's traffic but has no company location close to or connected to it.

The idea behind the PBX network is to permit any employee at any location to dial any other employee at any other location without paying the unit or long-distance charge normally required. In addition, each location is able to place a local call costing perhaps zero or one unit to any location where there is a company office or an FX line has been installed to a central office.

The savings in using dedicated lines in a centralized PBX network must, of course, be more than offset by the cost of all of the tie lines and any FX lines used. This is a function of the traffic volumes and the locations of company offices. It works best when the locations involved are spread rather far apart (i.e., do not share the same central offices) and each location is situated close to the source of some other location's business contacts. One negative factor that must be overcome by other advantages is that all calls must now travel through the centralized PBX location and take a longer path than would normally be required.

Voice Store-and-Forward Systems

Some PBX manufacturers are now offering a voice message storing system whereby a user can call in a regular phone conversation message and have it digitized and stored on a disk file for retrieval later by the called party. This is very similar to electronic mail, in which keyed-in data messages are stored on disk files for subsequent retrieval. However, there are some very distinct advantages over electronic mail: the spoken word is all that is needed, and data need not be keyed in on a typewriter or display. Also, special terminals are not required as with electronic mail; any push-button telephone instrument in the world can participate.

Each local phone station connected to the PBX has a *message waiting* lamp to notify a called party that a voice message is waiting to be retrieved *at your convenience*. Thus when a PBX phone station is busy, unattended, or in a do not disturb status, both outside and inside calls are routed to the disk file for storage. There is also the option to *broadcast* a single voice message to a fixed or variable distribution list of people.

To retrieve a message, the user calls the PBX, enters a password, and receives the message. This system ends the "telephone tag game" that takes up so much manpower and telephone facility capacity in today's business environment.

Military Connector Pin Standards

- MIL STD 188C is very similar to the EIA RS-232C standard.
- MIL STD 188-114 applies to a balanced circuit. It is the same as the EIA RS-422A standard except that less tolerance is allowed on one of the voltage levels.
- MIL STD 188-144 applies to an unbalanced circuit. It is the same as the EIA RS-423A except for the tolerance on a voltage level.
- MIL STD 1397 applies to speeds of 42,000 bps or 250,000 bps.

EIA Connector Pin Standards

- Pin 1, circuit AA, abbreviated name FG, is a protective frame ground.

- Pin 2, BA, TD, TRANSMIT DATA (signal direction is) to modem. Signals on this circuit are originated by the terminal and are transferred to the local signal converter for retransmission to the remote terminal. The circuit is kept in a MARK (binary 1) condition when data is not being transmitted. The following four circuits must have an ON condition for data to be transmitted: REQUEST TO SEND (CA), CLEAR TO SEND (CB), DATA SET READY (CC), and DATA TERMINAL READY (CD).

- Pin 3, BB, RD, RECEIVE DATA from terminal. Data signals received from the remote terminal are retransmitted to the local terminal. The circuit is kept in a MARK condition when circuit CF (RECEIVE LINE SIGNAL DETECTOR) is in an OFF condition. Under half duplex operation this circuit is held in a MARK condition when circuit CA (REQUEST TO SEND) is in the ON condition and briefly if circuit CA changes from ON to OFF to allow for completion of transmission.

- Pin 4, CA, RTS, REQUEST TO SEND from modem. This circuit controls the direction of data transmission. An ON condition sets this circuit in a transmit mode and an OFF condition in a receive mode. Under full duplex operation the ON condition keeps the modem on a transmit mode for long periods of time.

- Pin 5, CB, CTS, CLEAR TO SEND from modem. This circuit indicates whether or not the modem is ready to transmit data. An ON condition tells the terminal that signals sent to circuit BA (TRANSMIT DATA) will be sent over the communications line. Circuits CA, CC, and CD (if used) must also be in an ON condition. An OFF condition tells the terminal not to attempt to send data.

- Pin 6, CC, DSR, DATA SENT READY from modem. This circuit

indicates the status of the local modem. An ON condition tells the terminal that the modem is connected to the communications line, is not in the test, talk, or dial mode, and is ready to operate.

- Pin 7, AB, SG, SIGNAL GROUND, establishes a common ground for all circuits except AA, PROTECTIVE FRAME GROUND.

- Pin 8, CF, DCD, RECEIVE LINE SIGNAL DETECTOR from modem. The purpose of this circuit is to determine whether a suitable signal being received from the communications line can be demodulated and passed on to the terminal. An ON condition indicates yes; an OFF condition indicates no. For half duplex operation this circuit is held OFF whenever the REQUEST TO SEND is ON and briefly when it changes from ON to OFF.

- Pin 9 is used to test for a positive DC voltage.

- Pin 10 is used to test for a negative DC voltage.

- Pin 11 is used for the equalizer mode with Bell 208A modems.

- Pin 12, SCF, SDCD, SECONDARY RECEIVE LINE SIGNAL DE-TECTOR from modem. This circuit is comparable to pin 8, RE-CEIVE LINE SIGNAL DETECTOR, except that it applies only to the secondary channel if used. If the secondary channel is used for circuit assurance or as an interrupt channel, this circuit can indicate the status or signal the interrupt.

- Pin 13, SCB, SCTS, SECONDARY CLEAR TO SEND from modem. This circuit is comparable to pin 5, CLEAR TO SEND, except that it applies only to the secondary channel if used.

- Pin 14, SBA, STD, SECONDARY TRANSMIT DATA to modem. This circuit is comparable to pin 2, TRANSMIT DATA, except that it applies only to the secondary channel if used.

- Pin 15, DB, TC, TRANSMITTER CLOCK from modem. The signals on this circuit provide the terminal with signal element timing. The terminal must, in turn, provide a data timing signal on the TRANS-MIT DATA circuit BA whenever this circuit DB changes its signal from an OFF to an ON condition. This timing signal can be withheld by the modem for short periods if circuit CC, DATA SET READY, is in an OFF condition.

- Pin 16, SBB, SRD, SECONDARY RECEIVE DATA from modem. This circuit is comparable to pin 3, RECEIVE DATA, except that it applies only to the secondary channel if used.

- Pin 17, DD, RC, RECEIVE CLOCK from modem. The signals on this circuit provide the terminal with signal element timing from the remote location. This timing information shall be provided whenever circuit CF, RECEIVE LINE SIGNAL DETECTOR, is in the ON condition.

- Pin 18 is used for divided clock receiving with Bell 208A modems.
- Pin 19, SCA, SRTS, SECONDARY REQUEST TO SEND to modem. This circuit is comparable to pin 4, REQUEST TO SEND, except that it applies only to the secondary channel if used.
- Pin 20, CD, DTR, DATA TERMINAL READY to modem. This circuit controls the switching of the modem to the communications line. An ON condition prepares the modem to be connected to the line and will maintain the connection if made by external means such as manual call origination or answering or automatic call origination. If a station is in an automatic answer mode, a connection to the line will be made in the presence of a ringing signal and an ON condition of circuit CD, DATA TERMINAL READY.
- Pin 21, CG, SQ, SIGNAL QUALITY DETECTOR from modem. The signals on this circuit are employed to show whether the error rate of received data is excessive. An OFF condition indicates an alarm-type situation concerning an excessively high error rate.
- Pin 22, CE, RI, RING INDICATOR from modem. This circuit goes into an ON condition if a ring signal is received from the communications line.
- Pin 23, CH, DRS, DATA RATE SELECTOR to modem. The purpose of this circuit is to provide a means of automatically selecting a data rate between two predetermined speeds for synchronous modems or two ranges of speeds for asynchronous modems. An ON condition on this circuit will select the higher speed.
- Pin 24, DA, ETC., EXTERNAL TRANSMIT CLOCK to modem. This circuit is used to provide the timing signal from the terminal when that alternative has been selected and the terminal is in a POWER ON condition.
- Pin 25 is used to indicate a busy condition in which the modem will not accept incoming signals.

Other CCITT Interface Standards

- V.10 applies to unbalanced double-current interchange circuits with speeds up to 100,000 bps and is similar to the RS-423A standard.
- V.11 applies to balanced double-current interchange circuits operating at speeds up to 10,000,000 bps.
- V.21 applies to a full duplex switched network with speeds up to 300 bps, as with a Bell 103 modem.
- V.22 applies to a full duplex switched network with a speed of 1200 bps, as with a Bell 212A modem.
- V.23 applies to a half duplex switched network with speeds of 600 to 1200 bps, as with a Bell 202 modem.
- V.24 is the same as the EIA RS-232C standard.
- V.25 applies to automatic calling units, such as the Bell 801.
- V.26 applies to four-wire leased lines operation at 2400 bps, such as the Bell 201B modem, or to the switched network, as with the Bell 201C modem.
- V.27 applies to leased lines using manual or automatic equalizers at 4800 bps, such as the Bell 208A modem, or to a switched network at 2400 or 4800 bps.
- V.28 applies to unbalanced double-current interchange circuits at a speed of 20,000 bps.
- V.29 applies to circuits using an automatic adaptive equalizer and the QAM technique at a speed of 9600 bps.
- V.36 applies to synchronous data transmission at 48,000 to 72,000 bps using 60,000- to 108,000-bps group band circuits.
- V.57 deals with the standards for measuring equipment for data transmissions over 20,000,000 bps.

AT&T WATS Service

Wide Area Telecommunications Service (WATS) enables you to place (Outward) or receive (800 Service) toll calls over an access line within a specified calling period. The charge to install an Outward Wats access line is $174.80. The charge to install two 800 Service access lines is $366.95. There is a minimum requirement of two access lines per service group for 800 Service. The standard termination for all Interstate WATS access lines is the standard telephone jack. All prices quoted are for 1984.

Class of Service

Each access line is considered Measured Time Service. Usage for calls is billed according to the time of day calls are placed and at a decreasing rate as usage is increased.

Time-of-Day Rates

The 24-h day is split into three rate periods as follows:

Rate Period	Time of Day
Business day	8:00 A.M. to 5:00 P.M. Monday–Friday
Evening	5:00 P.M. to 11:00 P.M. Sunday–Friday
Night and weekend	11:00 P.M. to 8:00 A.M. Every day
	8:00 A.M. to 11:00 P.M. Saturdays
	8:00 A.M. to 5:00 P.M. Sundays

	SUN	MON	TUES	WED	THUR	FRI	SAT
8:00 A.M. to 5:00 P.M.	Night/ Weekend Rate	Business-day Rate (4 taper points)					
5:00 P.M. to 11:00 P.M.		Evening Rate (4 taper points) 35% discount (outward WATS) 28% discount (800 service)					
11:00 P.M. to 8:00 A.M.		Night/Weekend Rate (flat, no taper points) 65% discount (from highest day taper rate) 52% discount (from highest day taper rate)					

Tapered Rates

The charge for usage per hour decreases as the amount of usage increases. The taper points are as follows, and they apply only to business-day and evening rates:

0–15 h
15.1–40 h
40.1–80 h
80+ h
All night and weekend usage is billed at a flat hourly rate.
Usage is billed in $\frac{1}{10}$-h increments.

Table D-1   Interstate Outward WATS—Usage Rate Schedule.

Service Area	1–15 h		15.1–40 h		40.1–80 h		80.1+ h		Night and Weekend
	Business Day	Evening	Business Day	Evening	Business Day	Evening	Business Day	Evening	
1	$17.58	$11.43	$15.65	$10.18	$13.72	$ 8.92	$11.60	$ 7.53	$6.11
2	18.40	11.96	16.39	10.66	14.37	9.34	12.15	7.90	6.39
3	19.05	12.39	16.95	11.01	14.85	9.66	12.57	8.17	6.62
4	19.59	12.73	17.45	11.34	15.29	9.94	12.93	8.40	6.82
5	20.61	13.40	18.37	11.93	16.09	10.47	13.61	8.85	7.16
6	25.32	16.45	22.53	14.64	19.75	12.84	16.70	10.85	8.86

Table D-2 Interstate Inward WATS—Usage Rate Schedule.

Service Area	1–15 h		15.1–40 h		40.1–80 h		80.1+ h		Night and Weekend
	Business Day	Evening	Business Day	Evening	Business Day	Evening	Business Day	Evening	
1	$17.07	$12.29	$15.59	$11.22	$14.11	$10.17	$12.49	$ 9.00	$ 8.13
2	17.57	12.65	16.04	11.54	14.52	10.46	12.85	9.26	8.37
3	17.93	12.92	16.39	11.79	14.84	10.69	13.14	9.47	8.55
4	18.25	13.15	16.67	12.00	15.09	10.87	13.36	9.62	8.70
5	18.84	13.57	17.21	12.40	15.57	11.21	13.79	9.93	8.98
6	22.78	16.40	20.73	14.93	18.91	13.62	16.63	11.97	10.94

A Minimum Average Time Requirement (MATR) of 1 min per message will be applicable in each individual rate. If messages are greater than the minutes in any rate period, MATR applies, and chargeable minutes in that rate period will be equal to the number of messages in that rate period. MATR can apply in any or all of the three rate periods.

The usage rates are shown in Tables D-1 and D-2.

The monthly rate for Outward WATS is $31.65. The monthly rate for 800 Service (Inward WATS) is $73.60, which includes the minimum requirement of two lines. No usage allowance is included with the monthly rates.

Glossary

Access time—The time required to retrieve or store data that is on a file, such as a disk file.

ACF/VTAM—Advanced Communications Function for the Virtual Telecommunications Access Method. A set of software instructions that control the communications of a host computer compatible with IBM's Systems Network Architecture.

Acoustical coupler—A device that permits the use of a regular telephone handset to transmit data by converting the terminal's digital signals to audio signals.

ACU—*See* Automatic calling unit.

Adaptive differential pulse code modulation—A CCITT standard encoding technique permitting an analog voice conversation to be carried within a 32,000 bps digital channel.

Address—Coded characters that identify a terminal, a location in storage, a destination, etc.

ADPCM—*See* adaptive differential pulse code modulation.

Algorithm—A set of rules and mathematical calculations used in a computer program to solve a problem or obtain a desired factor, such as a random number.

Alphanumeric—An alphabetic letter, numeric digit, or special symbol.

Alternate routing—A secondary communications path to be used if the primary path becomes unavailable.

ALU—Arithmetic and logic unit.

Amplification—The strengthening of a weak signal so that it can be transmitted a greater distance.

Amplitude—The maximum variation of a waveform from its zero value.

Amplitude modulation—The variation of the amplitude of a sine-wave signal to provide it with intelligence, such as the letter A.

Analog—A signal that is varied continuously by some physical factor, but not discretely with identifiable pulses as with digital signals. It is the method originally used to transmit voice signals over the telephone network.

ANSI—American National Standards Institute. An organization formed in the United States to establish telecommunications standards.

Answerback—A coded signal emitted by a terminal in response to a received signal from another terminal (or computer) indicating that it has identified itself and is ready to receive data.

Application program—A set of software instructions that are entered into a computer as part of a business system, such as a payroll check printing program.

ARQ—Automatic request to repeat. A request to the transmitting station to resend a block of data that was received in error.

ARS—*See* Automatic route selection.

ASCII—American Standard Code for Information Exchange. An eight-bit code that was established by the American National Standards Association.

ASR—Automatic send and receive. The ability of a teletype-like terminal to transmit and receive data automatically (rather than have each character keyed individually) through the use of a cassette or some data storage device.

Asynchronous—A method of transmitting data whereby each character is identified by start and stop bits, rather than by a timing procedure.

Attended operation—A data station that requires the intervention of an individual to complete a transmission, such as the depression of a push button to put a modem in the DATA mode.

Attenuation—The reduction in power (current or voltage) of a signal as it travels from one location to another. It is usually expressed in decibels.

Audio frequency—A transmission frequency that is audible to the human ear, usually in the range of 30 to 20,000 Hz.

Auto-answer—Unattended operation of a modem, terminal, etc. whereby dial-up calls are answered automatically to establish a data connection with a remote device.

Autodial—*See* automatic calling unit.

Automatic calling unit—A device that allows a terminal or computer to dial a telephone automatically, without human intervention.

Automatic route selection—The capability of a switch (PBX) to automatically determine the most desirable route for establishing a connection with a remote device.

BAL—Basic assembly language. A set of IBM computer programming instructions used to create operational computer programs.

Bandwidth—The range of frequencies that can be transmitted over a particular communications channel, which also determines the maximum bits per second that the channel will accommodate without excessive distortion.

Baseband—A basic signal transmitted at its original frequency without external modulation, such as a digital signal.

Batch processing—The processing of large quantities of similar data on a computer at one time rather than interactively or in real time.

Baud—A measure of transmission speed based upon the number of signal conditions per second rather than the types of signals. If there is only one type (level) of signal, the baud equals bits per second.

Baudot code—An older data transmission code, currently used in Telex, in which five bits define each of 32 alphanumeric characters.

BCC—*See* Block check character.

BCD—Binary-coded decimal. A six-bit code not in popular use today.

BERT—Bit error rate test. A standard procedure used for the comparison of the percentage of erroneous data bits received.

Binary—A numbering system using a base of 2, rather than 10. The digits are 0 and 1.

Bipolar—A method of manufacturing integrated circuits by using layers of silicon with different electrical characteristics. It is also a method of coding in which a 0 is transmitted as no pulse and a 1 as a pulse.

Bisynch—Binary synchronous communications (BSC). A communications protocol with a discipline of control characters for synchronous transmissions.

Bit—A binary digit which can be a 0 (space) or a 1 (mark).

Bit rate—The speed at which data is being transmitted; it is stated as the number of bits per second (bps).

Bit stuffing—The insertion of an extra bit in a data stream so that the data rate is less than the channel bit signaling rate.

BIU—Basic information unit. The smallest frame of data transmitted within IBM's Systems Network Architecture.

Block—A group of data characters handled as a discrete entity by its size or by its starting and ending control character delimiters.

Block check character—A control character that is appended to a block of data and is used to determine if the block was received in error, via longitudinal or cyclic error checking calculations.

BLU—Basic link unit. The largest frame of data transmitted within IBM's Systems Network Architecture.

BOC—Bell Operating Company is one of the current 22 local telephone companies previously associated with AT&T.

bps—*See* Bit rate.

Broadband—A communications channel with a large bandwidth, certainly greater than that of a voice-grade channel.

BTAM—Basic Telecommunications Access Method. A computer program offered by IBM to control the communications by a host computer.

Buffer—A finite memory area in which data is stored temporarily to compensate for a data rate or time availability difference between two devices such as a computer and a terminal.

Bus—An electrical connection of one or more conductors creating a transmission path or channel where all attached devices receive all of the transmissions at the same time and each device has to select its designated portion.

Busy signal—An audible or flashing signal indicating that a number called is not available.

Byte—A group of adjacent bits, normally eight, handled as a unit of information.

Byte multiplexing—An option in time-division multiplexing whereby the transmission time slots are based on a byte rather than a bit.

Byte stuffing—Insertion of an extra byte in a data stream so the data rate is less than the channel signaling rate.

CAD/CAM—Computer-aided design and computer-aided manufacturing.

Call detail recording—A PBX feature whereby phone calls are logged, usually by time, destination, and charges, for subsequent use in establishing charges by department, etc.

Camp on—A telephone call to a busy number that results in a wait condition. Each party receives a ring when the called number becomes available.

Carrier—A signal with a constant frequency which can be modulated with a second, information-type signal.

Carrier system—A communications line that accepts different carrier frequencies that can be modulated to create multiple communications channels.

Cassette—A magnetic tape used to store data or messages for subsequent transmission or processing.

Cathode ray tube—An electronic screen, similar to a TV tube, that is used to display data.

CCITT—International Telegraph and Telephone Consultive Committee. A worldwide organization dealing with standards for telephone, telegraph, and data transmission.

CCS—A unit of measure, consisting of 100 calls per second, used in telephone line traffic loading evaluation.

CCSA—Common control switching arrangement. Leased line telephone switching facilities, controlled by an individual business, whereby many scattered locations can be connected to each other without using the telephone company's toll facilities.

CDMA—*See* Code-division multiple access.

CDR—*See* Call detail recording.

Central office—A telephone company switching center that accommodates local area subscribers and lines for switching to other areas and their subscribers.

Central processing unit—The microprocessor portion of a computer that executes the coded instructions; often used to refer to the computer itself.

Centrex—A telephone switching system, offered by the Bell System, in which the switch, or PBX, is normally located at the phone company central office.

Channel—A single communications path. It could be one of several available paths in a physical communications line.

Channel service unit—A device used to terminate a digital circuit at the customer's premises; it performs line conditioning functions, loopback tests, etc.

Chip—An integrated circuit of silicon construction that is fabricated to contain numerous semiconductor circuits.

CICS—Customer Information Control System. An IBM computer program that controls the communications between a terminal operator and the application program in the host computer. It handles on-line files and data communications.

Circuit—A facility for providing communications between two or more points by using any of the communications media such as two-wire, four-wire, and microwave.

Clock—An electrical circuit that generates precisely timed signals to control synchronous transmission timing.

Cluster controller—A device like the IBM 3724 that provides remote data communications for a group of CRT displays and printers.

CMOS—Complementary metal-oxide semiconductor.

Coaxial cable—A cable used for higher-bandwidth communications. It consists of a cylindrical outer conductor from which a single-wire inner conductor is separated by an insulated spacer.

COBOL—Common business oriented language. A programming language written for the business field.

Code-division multiple access (CDMA)—A method by which multiple users can access a satellite. Each user has a unique code.

Common carrier—A company that is authorized to provide communications transmission facilities to the public. The Bell System is an example.

Compandor—A device used to improve the signal-to-noise ratio of a communications link by first compressing the volume range of signals and later expanding them.

Compression—Reduction of the number of bits required to represent information during transmission or in storage (e.g., use 80 B instead of 80 blanks).

COMSAT—Communications Satellite Corp.

Concentrator—An electronic device that connects multiple lower-speed lines that are used intermittently to lower number higher-speed lines for a more economical total transmission path.

Conditioning—The addition of electronic components to a communications channel to improve transmission capability by reducing such impediments as delay distortion and attenuation distortion.

Contention—Competition between multiple devices to utilize the same facility, such as two terminals attempting to transmit on the same communications channel at the same time.

CPE—*See* Customer premises equipment.

Conversational—Time dependent data transmission where an operator enters data and waits for a response before continuing (also known as interactive).

CPU—*See* Central processing unit.

CRC—*See* Cyclic redundancy check.

Crossbar switch—An older type of telephone switching mechanism that can connect any combination of multiple vertical and horizontal paths.

Cross talk—The inadvertent mixing of signals between two closely associated communications channels.

CRT—*See* Cathode ray tube.

CSMA/CD—Carrier sense multiple access with collision detection is a local network access control procedure used to prevent interference between multiple transmitting stations by having them sense an occupied circuit prior to transmission.

CSU—*See* Channel service unit.

CTS—Clear to send.

Customer premises equipment—Equipment, such as a PBX, that is installed at a user's location and connects to the telephone network.

Cyclic redundancy check—A method of error checking with a block or frame of data whereby a mathematical calculation is performed at each end and compared for accuracy.

DAA—*See* Data Access Arrangement.

Data—Information in the form of numbers, text, etc.

Data Access Arrangement (DAA)—An electronic circuit box supplied by the Bell System to prevent undesirable signals from entering the switched network.

Data communications—The transmission of information between separate terminals or computers.

Data communications equipment—The device (modem) that provides the signaling interface between a terminal and a communications channel.

Data encryption—The manipulation of information during transmission to prevent its use by anyone other than those stations that have been provided with the decoding key.

Dataphone—A modem supplied by the Bell System.

Dataphone Digital Service (DDS)—Communications channels offered by the Bell System and AT&T that accept digital, rather than analog, signals.

Data Service Unit (DSU)—A replacement for a modem in AT&T's Dataphone Digital Service offering.

Data set—A term used by the Bell System and AT&T to describe a modem.

Data terminal equipment—The terminal that is normally connected to a modem and which supplies the data for transmission.

DCE—*See* Data communications equipment.

DDD—*See* Direct distance dialing.

DDS—*See* Dataphone Digital Service.

Decibel (dB)—A unit of measure of the strength of a signal. More specifically, the transmission (power) level at one point on a circuit as compared to another point (zero level).

Dedicated line—A communications line that is assigned to a single subscriber. Also termed *private line* and *leased line*.

Delay distortion—Distortion caused by a signal being nonlinear over a range of frequencies.

Demodulation—The separation of an intelligent waveform from its nonintelligent carrier signal, the reverse of modulation.

Dial tone—A 90-Hz telephone signal that is sent to a user as notification that dialing can commence.

Dial-up—The process of initiating a connection on the telephone company's switched network. A dial-up line is a switched network line.

Digital—A signal that is varied discretely with identifiable pulses, such as a square wave normally used to transmit data.

Direct distance dialing (DDD)—A service offered by the telephone company that permits the user to dial the call without operator assistance. Also, the use of regular dial-up service rather than the alternatives of private lines, WATS, etc.

Direct inward dialing—A feature of the Bell System's Centrex PBX switching arrangement that permits outside calls to be connected directly to the called party without going through a company operator.

Distributive data processing—A network of geographically dispersed computers or data processing stations that share a common resource such as a database or application programs often residing in a large host computer.

DLC—Data link control; also called *communications line discipline.*

DOS—Disk operating system.

DS-1—Digital signal level 1 identifies the 1.544 Mbps digital signal carried in a T1 channel facility (DS-0 = 64,000 bps, DS-1C = 3.152 Mbps, DS-2 = 6.312 Mbps).

DSU—*See* Data Service Unit.

DTE—*See* Data terminal equipment.

DTMF—*See* Dual tone multifrequency.

Dual tone multifrequency—The touch-tone signaling procedure in which each depression of a key generates two audible tones that combine to represent one of the 12 telephone keys.

Duplex transmission—Also called full duplex. Simultaneous two-way transmission.

EBCDIC—Extended Binary Coded Decimal Interchange Code. The code originated by IBM for use in its computer systems to identify information.

Echo—A signal wave that has been reflected back to the source. If it is of sufficient magnitude, it is troublesome.

Echo suppressor—An electronic device used to combat echoes in long-distance telephone connections.

EDP—Electronic data processing.

EIA—Electronic Industries Association. A group that provides standards for interface voltages between terminals and modems.

Encode—To convert characters into their equivalent bit structures.

ENQ—Enquiry in bisynch protocol.

EOA—End of address in bisynch protocol.

EOM—End of message in bisynch protocol.

EOT—End of transmission in bisynch protocol.

EPROM—Erasable programmable read-only memory.

EPSCS—Enhanced Private Switched Communication Service (AT&T).

Equalization—An adjustment to a communications channel characteristic to compensate for signal loss and delay.

Erlang—A measurement of communications line traffic intensity used in line-loading analysis.

ESS—Electronic switching system. A telephone company switching computer, such as the Bell System's ESS 1.

ETX—End of text in bisynch protocol.

Exchange—A telephone switching center employed by a common communications carrier such as the Bell System.

Expander—An electronic device that produces a larger range of output voltages for a given input voltage.

Facsimile—A system used to transmit images (letters, pictures, etc.) by scanning the original documents.

FCC—*See* Federal Communications Commission.

FDM—*See* Frequency-division multiplexing.

FDMA—*See* Frequency-division multiple access.

FDX—Full duplex; *see* Duplex.

FEC—*See* Forward error correction.

Federal Communications Commission (FCC)—A board, consisting of seven commissioners, that regulates all interstate and overseas communications systems.

FEP—*See* Front end processor.

Fiber optics—Waveguides constructed of fine strands of glass that are used to transmit signals similarly to a telephone line. The source of energy is a light-emitting diode (LED) or a laser beam.

Frame—A unit of data transmission, similar to a block, which is made up of a set of consecutive time slots identified by a frame-alignment signal.

FM—*See* Frequency modulation.

Foreign exchange line (FX)—A dedicated line connecting a user's telephone station to a distant telephone exchange to take advantage of local service rates at that exchange.

Forward error correction—A method of error detection and correction whereby an algorithm is applied to add extra bits to a data stream which can be used at the receiving end to reconstruct any incorrectly received data.

Four-wire circuit—A communications circuit utilizing two pairs of conductors—two to transmit and two to receive signals.

Frequency—The rate at which a signal current alternates as measured in hertz, or cycles per second.

Frequency band—A range of frequencies, such as 300 to 3400 Hz, that can be accommodated on a single channel.

Frequency-division multiple access (FDMA)—An arrangement whereby multiple devices can share a common communications link by the assignment of different frequencies.

Frequency-division multiplexing (FDM)—The separation of a communications link into multiple traffic channels by the assignment of separate frequencies.

Frequency modulation (FM)—The process of modifying a carrier sine wave to include intelligence by changing the frequency at discrete points in time.

Frequency shift keying (FSK)—The process of modifying a carrier sine wave to include intelligence by changing the phase of the sine wave at discrete points in time.

Front end processor (FEP)—A communications computer that is situated in front of a mainframe computer to handle the data communications lines.

FSK—*See* Frequency shift keying.

FX—*See* Foreign exchange line.

Full duplex (FDX)—*See* Duplex.

Gain—An increase in output signal level (power) over that of the input signal level, usually caused by an amplifier.

Gateway—A network station performing protocol conversion to connect two otherwise incompatible communications networks.

Geosynchronous orbit—A satellite whose orbit matches the rotation of the earth so the satellite remains relatively stationary to a preselected point on the earth.

Half duplex (HDX)—A communications channel over which signals can be transmitted in only one direction at any one time and so must be "turned around" to permit the receiver to become the sender.

Handshaking—The exchange of predetermined signals by two communicating devices, such as modems, to establish control of a new connection.

Hard copy—A term used to indicate printed paper information as compared to information on disk files, etc.

Hard-wired—A connection (modem to line, etc), using the physical wire rather than something like an acoustical coupler.

Harmonic distortion—Signal distortion caused by nonlinear characteristics of a transmission channel that creates undesirable harmonic frequencies.

HDLC—High-Level Data Link Control. Established by ISO for bit-oriented data link protocol.

HDX—*See* Half duplex.

Header—The control information appearing at the beginning of a transmission message to indicate such things as source, destination, and priority.

Hexadecimal—A numbering system using a base of 16, rather than 10, which is widely employed in computers.

Hookswitch—A switch located in a telephone instrument cradle which activates the electric circuit when the handset is lifted.

Host computer—A computer, attached to a communications network, that provides the main source of file information and control of the operation of the network.

IC—Integrated circuit. A semiconductor that provides one complete electric circuit.

IEEE—Institute of Electrical and Electronics Engineers.

IMS/VS—Information Management System/Virtual Storage. An IBM program that runs in a host computer and is responsible for accepting transactions from a terminal, processing them, accessing databases, and returning the output to the terminal.

INTELSAT—International Telecommunications Satellite Consortium. An organization formed in 1964 to establish a worldwide satellite communications system.

Interactive—*See* Conversational.

Interface—A physical point of demarcation between two unlike devices where certain procedures, codes, protocols, etc. must be pre-defined to permit the transfer of information.

International Telecommunications Union (ITU)—An organization formed to standardize communications procedures around the world as an agency of the United Nations.

Interoffice trunk—A communications line connecting telephone company local central offices.

I/O—Input/output.

I/O channel—The portion of a computer that controls the transfer of data between the mainframe and its peripherals (disk drives, FEP, etc.).

ISDN—Integrated Services Digital Network. A proposed standard universal public network for data, voice, facsimile, and video communications.

ISN—Information Systems Network, AT&T's LAN.

ISO—International Standards Organization. An organization formed to establish standards for worldwide communications procedures.

Isochronous transmission—A combination of synchronous and asynchronous methods of communications in which clocking is employed but the data is also framed with start and stop bits.

ITB—Intermediate text block in bisynch protocol.

ITU—*See* International Telecommunications Union.

JCL—Job control language. A set of computer instructions and procedures employed by computer programming personnel to control the method of executing computer application programs.

kbps—kilo (thousand) bits per second.

Keyboard-send-receive (KSR)—The ability of a teletype-like terminal to transmit data only via the keyboard at a rate lower than by automatic transmission (ASR).

Key telephone—A multiple push-button type telephone instrument that can be connected to multiple telephone lines and can place lines on hold.

LAN—Local area network. The integration of communications lines within a single building or company complex.

Laser—A device that emits a narrow beam of electromagnetic energy as used in POS system scanners.

LATA—*See* Local access and transport area.

Leased line—A private communications channel the use of which is restricted to the subscribing company. Also called a *dedicated line.*

Least cost routing—*See* Automatic route selection.

LED—Light-emitting diode. An electronic device that emits a small amount of light and consumes little current. It is installed in modems and other electronic devices.

Level—The amplitude or strength of a communications signal.

Line printer—A higher-speed printer that prints an entire line at a time in contrast to the lower-speed character printer.

Link (communications)—A communications channel that connects two stations to each other.

Loading coil—An induction coil of wire used in local telephone loops exceeding 18,000 ft to compensate for the wire's capacitance and to boost the voice-grade frequencies.

Local access and transport area—The United States has been divided into 161 local telephone company geographical areas which are serviced by the Bell Operating Companies.

Local loop—The two- or four-wire line connection that ties a user's location to the nearest telephone company central office.

Longitudinal redundancy check (LRC)—A method of transmission error detection that employs a parity check of each of the characters in a block of data.

Loopback—A procedure for testing communications lines whereby the original transmitted signal is switched back to its origin, where its integrity is verified.

LRC—*See* Longitudinal redundancy check.

LSI—Large-scale integration. A small silicon or integrated circuit that contains many separate circuits, as used in today's microprocessors.

LU—Logical unit. A term used in IBM's Systems Network Architecture to describe a port through which a user accesses a communications system.

Mark—Originally the presence of a signal in telegraph communications. Now used to designate a binary 1.

Mark hold—The continuous transmission of mark pulses to signify a nontraffic line condition.

Message switching—A system of relaying messages that are transmitted to one location, perhaps stored temporarily, and then retransmitted to another location, all of this usually being performed automatically by a computer.

MATR—Minimum Average Time Requirement (AT&T).

Microcomputer—A category of smaller computers in which a single circuit board contains all of the central processor functions.

Microprocessor—A single LSI (large-scale integration) chip.

Microwave—The transmission of radio-frequency signals as a communications medium. Dish antennas rather than physical communications lines are used.

Minicomputer—A category of small computers that usually handle 8- to 16-bit words. (There are super-minicomputers that handle 32-bit words.)

MIPS—Million instructions per second is a general comparison gauge of a computer's processing speed and power.

MNP—A protocol introduced by Microcom and offered by Telenet and Uninet packet switching networks.

Modem—Electronic equipment that modulates digital signals and demodulates analog signals so the digital signals can be transported over an analog telephone line.

Modulation—A predefined change in the amplitude, frequency, or phase of an analog telephone signal to define digital data.

MOS—Metal oxide semiconductor. A type of large-scale integration.

MSNF—Multi System Networking Facility (IBM).

Multidrop line—A circuit or line connecting multiple locations that share its use.

Multiplexing—An electronic device that provides for the transmission of multiple separate signals over a common physical communications link.

NAK—Negative acknowledgment in bisynch protocol.

NAU—Network addressable unit. A term used in IBM's Systems Network Architecture to describe the elements of the system that can be addressed (i.e., SSCP, PU, and LU).

NCP—Network Control Program. A set of instructions that reside in the front end processor and control the communications traffic.

Network—A group of scattered stations, such as the telephone company's switched network or a private network, that are connected together by communications channels so they can transmit to each other.

Node—The end point of a data communications link or a junction common to multiple links in a network.

Noise—Undesirable electrical signals that degrade the performance of a communications channel.

NRZ—Non-return to zero concerns pulse codes where no additional transitions are induced and one signal level indicates one logic state and another signal level indicates the other logic state.

NRZI—Non-return to zero inverted concerns pulse codes on which a change in signal level denotes a logic 0 and no change denotes a logic 1.

OCR—*See* Optical character recognition.

Off-hook—A telephone handset that has been placed in operation, as by manually lifting it from the cradle.

Off-line—A device not connected to a computer or a communications line.

On-line—A device connected to a computer or a communications line.

Optical character recognition—The process of reading graphic characters with a light-sensitive device which converts them into machine-readable digital codes.

Operating system—The software in a computer that controls the execution of programs, handling such functions as input/output control, resource scheduling, and data management.

Optical fiber—A filament of fiber that is used to transmit laser or LED-generated light signals as a wave guide.

OS/VS, OS/MVS—Operating system virtual storage and multiple virtual storage. Sets of programmed instructions that manage all of the available resources of a host computer.

Packet—A group of binary digits (often 128), including control information and data, that is transmitted as an independent block of information.

Packet switching—A system of relaying packets of information that are transmitted to one location, perhaps stored temporarily, and then retransmitted to another location, all of this usually being performed by computers.

PAD—Packet assembler/disassembler. A microprocessor-controlled box that performs the protocol conversion between a packet switching network and a nonpacket terminal.

Parallel transmission—The simultaneous transmission of all of the bits that comprise a character by utilizing multiple communications channels.

Parity check—A method of error detection whereby the 1 bits that make up a character (or a block of data) are summed and a bit is added to make all transmissions either even or odd in total parity. Vertical parity refers to the count for a single character; horizontal parity refers to the total count for a block of data.

PBX—Private branch exchange. A device that provides for the switching of local telephone stations to connect them to the public network.

PCM—*See* Pulse code modulation.

Phase distortion—Distortion caused by a difference between maximum and minimum transmission time of frequencies within a specific band. Also referred to as *delay distortion*.

Phase jitter—An undesirable variation in the timing of a signal.

Phase modulation—The process of modifying a carrier sine wave to include intelligence by changing the phase of the sine wave. Also referred to as *phase shift keying*.

Phase shift keying—*See* Phase modulation.

PIU—Path information unit. An intermediate size frame of data transmitted within IBM's Systems Network Architecture.

Polling—A method of controlling the transmissions of multiple stations on a communications line by inviting each station by address to transmit data in its turn.

Port—A point of access into a computer, network, or other electronic device.

POS—Point of sale. A computer business system in which terminals are installed in retail stores to record transactions as they occur.

Private line—A communications channel dedicated to predetermined groups of using stations. Also called a *leased line*.

PROM—Programmable read-only memory. A set of programmed instructions that cannot be written over and lost.

Propagation delay—The time it takes for the transmission of a signal from one point on a circuit to another point.

Protocol—A set of formal rules for regulating communications between two or more data stations.

PSK—Phase shift keying; *see* Phase modulation.

PU—Physical unit. A term used in IBM's Systems Network Architecture to describe an actual physical node or data station.

Pulse—A momentary change of finite duration in the current or voltage on a circuit.

Pulse code modulation—The conversion of an analog speech signal to an equivalent binary character. It is often referred to as digitizing a voice signal.

Quadrative amplitude modulation (QAM)—A means of increasing channel capacity by a combination of phase shift and amplitude modification.

Queue—A holding area for data or messages waiting their turn to be processed.

Radio wave—Electromagnetic signals of the approximate frequency range of 10,000 Hz to 3,000,000 MHz.

RAM—Random-access memory. A means of storing information whereby any portion of the entire file can be accessed directly without having to read the unwanted portions.

Random noise—Random disturbances that occur on a communications line. They are due to unpredictable electronic malfunctions.

Read—To extract information from the memory of a computer, disk file, etc.

Real time—The time at which information is generated rather a later, more convenient time for processing it.

Redundancy—The duplication of information or equipment as backup to ensure the operation of a critical facility. For example, data is transmitted twice to be sure that it is received, or an extra modem or computer is supplied as a backup for a critical component in a system.

Refresh rate—The rate at which an image on a CRT display screen is renewed in order to appear stable (e.g., 60 times per second in the United States).

Register—A device, often a microprocessor, that stores a single word of data (e.g., 32 bits).

Repeater—An electronic device used on communications lines to receive signals and retime and restrengthen them prior to retransmission to their destinations.

Response time—Generally, the time required for a device to respond to an input signal. Specifically, the time lapse from when a data entry operator presses a SEND key to when the requested data or an acknowledgment is received from the associated computer. Sometimes the time cycle is taken to end when the first character strikes the operator's screen, but a more practical approach might be to end the cycle when sufficient data has been received on the screen to release the keyboard for further use.

Reverse channel—*See* Secondary channel.

RF—Radio frequency.

RJE—Remote job entry. A line printer type of terminal that is used to initiate jobs to be run on a host computer.

ROM—Read-only memory. Programmed instructions that cannot be destroyed by inadvertently writing over them.

RTS—Request to send.

RZ—Return to zero concerns pulse codes in which a change in signal level denotes a logic 1 (mark) and no change denotes a logic 0 (space).

SBS—Satellite Business Systems (IBM).

Scrambling—A method of coding digital signals to maintain a more constant power level by producing a random signal pattern.

SDLC—Synchronous Data Link Control. IBM's bit-oriented protocol.

Secondary channel—A second channel made available on the same communications line as the primary channel. It can be used for data transmission at a lower rate or for control signals. An example is an audible tone sent back to the transmitting source to acknowledge a message.

Serial transmission—The transmission of each bit that makes up a character one at a time on a single channel. See also *parallel transmission*.

Sideband—A secondary channel that is created at the upper or lower side of the main carrier frequency and is used for low-speed transmissions such as modem diagnostic signal systems.

Signal—An electrical wave that is transmitted along a communications channel.

Signal-to-noise ratio—The ratio of usable signal strength to the undesirable communications channel noise. It is expressed in decibels.

Simplex—A communications channel that permits transmission in only one direction without the option to turn the line around and transmit in the other direction as with half duplex.

SNA—Systems Network Architecture. IBM's overall planned total system structure for a communications network.

Software—Computer programmed instructions, as compared to the hardware.

SOH—Start of header in bisynch protocol.

Solid state—Semiconductor components of electronic devices, such as computers and modems, that control the flow of electrons by means of materials such as silicon.

Space—A signal pulse representing a binary 0, as opposed to a mark or binary 1.

SPRINT—GTE Sprint Communications' public microwave network alternative to direct distance dialing.

SSCP—System service control point. A term used in IBM's Systems Network Architecture to describe a program that resides in the host computer to handle operator requests, establish sessions, etc.

Start-stop transmission—Asynchronous transmission where each character is preceeded by a start bit and followed by a stop bit.

Station—Any location containing a data terminal, computer, telephone, facsimile, or any other device that can participate in a communications activity.

Store and forward—A communications system in which information is received and stored at intermediate nodes for subsequent transmission to the final destination.

STX—Start of text in bisynch protocol.

Switched line—A communications link where the physical path is established by dialing, and varies with each use.

Synchronous—A communications method whereby the transmitting and receiving locations operate on a series of timing pulses to identify characters, etc.

Tandem office—A telephone company switching center connecting two local central offices.

Tariff—A federally regulated contractual agreement between the common carrier and the customer concerning use of a telephone company's facilities.

TASI—*See* Time assignment speech interpolation.

TCAM—Tele-Communications Access Method. An IBM set of programmed instructions used to control the communications for a host computer.

T-Carrier (T1, T1C, T2)—A time division multiplexed digital transmission facility operating at 1.544 Mbps or above.

TDM—*See* Time-division multiplexing.

TDMA—*See* Time-division multiple access.

Telecommunications—The transmission of information (data, audible signals, pictures, etc.) via any communications medium (phone line, microwave, satellite, etc.).

Teleprinter—An asynchronous start/stop communications terminal containing a keyboard and a printing mechanism. Also called a *teletypewriter*.

Teletype—An offering of teletypewriters by the Teletype Corp.

Teleprocessing—The transmission of computer data excluding voice conversations.

Telex—A worldwide service for teleprinters (and computers) utilizing the Baudot code.

Telpak—A communications offering of the Bell System and other common carriers concerning the leasing of groups of 60 or 240 voice-grade channels at a discounted rate.

Terminal—An electromechanical device used to send, receive, or simply create information. It may include a keyboard, printer, CRT screen, disk file, etc. A computer can be considered a terminal in the general sense.

Text—The information portion of a data transmission as compared to the header and trailer control data.

Throughput—The total quantity of useful information that can be passed through a communications facility in a specified period of time.

Tie line—A leased line connecting two or more points together. It is usually used to connect telephone switchboards or PBXs.

Time assignment speech interpolation—A method of transmitting multiple voice conversations on the same communications channel during lull periods.

Time-division multiplexing—The sharing of a faster communications line by multiple slower stations based upon alternate use of time slots.

Time-division multiple access (TDMA)—An arrangement whereby multiple devices can share a common communications link by allocating time slots to the various devices.

Timesharing—A method of making the facilities of a computer available to multiple users, each of which takes a turn at executing a limited number of instructions. Because of its speed, the computer normally appears to be completely available to each of the sharing users.

Token bus—A local area network access procedure where all stations attached to the bus have to sense a token or supervisory frame which will allow them to transmit data.

Toll center—A telephone company switching center for dial-up lines.

Touchtone—A Bell System term for push-button dialing as contrasted to rotary dialing.

Transceiver—A terminal that can both transmit and receive information.

Transistor—A chip of silicon material that switches or amplifies an electric current.

Translate—The process of converting information from one code to another (e.g., ASCII to EBCDIC for computer use).

Transparency—Data that can pass through a communications network without being changed or acted upon like a control code. Computer programs must be transmitted to another location in the "transparent mode," or they may trigger some control code functions inadvertently.

Transponder—A satellite component that receives signals from earth stations at one frequency, amplifies them, and retransmits them to other earth stations at a higher frequency.

Trunk—Telephone company circuits that connect switching centers.

TSO—Time Sharing Option. A set of programmed instructions offered by IBM for the development of business application programs on a host computer.

TTL—Transistor-to-transistor logic.

TTY—Teletypewriter equipment.

Turnaround time—The time required to reverse the direction of transmission from send to receive or vice versa on a half-duplex channel.

Two-wire circuit—A communications line made up of two-wire conductors that are insulated from each other and not grounded.

TWX—Teletypewriter Exchange Service. A public teletype network operated in the United States by Western Union.

Unattended operation—A data station that can be operated without need for human intervention, as by pushing a button to go to DATA on a modem.

Unipolar—Integrated circuit transistors so constructed that the flow of current is positive or negative but not both.

Unix—Bell Telephone Laboratories designed computer operating system for telecommunications and multiuser environments.

Value-added network (VAN)—A common carrier offering that provides a message or packet switching facility plus the possibility of information storage or even information processing.

Vertical redundancy check—An odd or even parity check performed on each character on a received block of data using the ASCII code.

Videotex—An interactive data communications application with a remote database for use by unsophisticated subscribers using television sets or some other low cost terminal.

Video signals—Signals, with frequencies up to 10,000,000 Hz, that are used to transmit pictures, such as television.

Virtual circuit—A circuit that is not a direct physical connection of one point to the other but is instead accomplished by an alternate means.

Voice digitation—Conversion of analog voice signals into digital signals for transmission or storage.

Voice-grade channel—A communications channel that is restricted to the normal voice frequencies of 300 to 3400 Hz, although it can be used for voice, data, facsimile, etc.

Volatile memory—An information storage facility (microprocessor, etc.) that loses its contents if the power source is removed.

VRC—*See* Vertical redundancy check.

VTAM—Virtual Telecommunications Access Method. An IBM teleprocessing program.

WATS—Wide Area Telephone Service. A telephone company offering involving an hourly charge for various calling zones in the United States. The switched telephone network is used.

Waveguide—Metallic tubing used for the directing of microwave transmission signals.

Wideband—A communications channel providing a bandwidth greater than a voice-grade channel, usually above 9600 bps.

Word—A group of bits (4, 8, 16, or 32) that are moved and stored as a unit.

Write—To load information into the memory of a computer, disk file, etc.

XNA—Extended Network Architecture.

X-Off/X-On—Transmitter OFF and ON codes are sent to prevent the overflow at a slower receiving device like a teletypewriter.

Index

267